REDBACK 12

CROSSING THE LINE

AUSTRALIA'S SECRET HISTORY IN THE TIMOR SEA

Kim McGrath

Published by Redback Quarterly,
an imprint of Schwartz Publishing Pty Ltd
Level 1, 221 Drummond Street
Carlton VIC 3053, Australia
enquiries@blackincbooks.com
www.blackincbooks.com.au

Copyright © Kim McGrath 2017
Kim McGrath asserts her right to be known as the author of this work.

ALL RIGHTS RESERVED.

No part of this publication may be reproduced, stored in a retrieval system, or transmitted in any form by any means electronic, mechanical, photocopying, recording or otherwise without the prior consent of the publishers.

National Library of Australia Cataloguing-in-Publication entry:
McGrath, Kim, author.
Crossing the line: Australia's secret history in the Timor Sea/Kim McGrath.
9781863959360 (paperback)
9781925435740 (ebook)
Offshore gas industry – Timor Sea.
Offshore gas industry – Moral and ethical aspects – Australia.
Offshore oil industry – Timor Sea.
Offshore oil industry–Moral and ethical aspects – Australia.
Timor-Leste – Relations – Australia.
Australia – Relations – Timor-Leste.

Cover design by Peter Long
Text design and typesetting by Peter Long
Cover image first published in the *Reader's Digest Great World Atlas*, 1962
© Harper Collins; maps courtesy of the Timor-Leste Maritime Boundary Office; file on p. 105 – NAA: A10463, 801/13/11/10 Part 2; photos on p. 111 – NAA: A10463, 801/13/11/3 Part 5; *Sydney Morning Herald* article on p. 143 – Fairfax Syndication; photo on p. 150 – Lyndon Mechielsen/Newspix; file on p. 197 – NAA: A1838, 1733/3/2 Part 7

For my parents, Margaret and Kevin

CONTENTS

Introduction

1

CHAPTER 1

Australia Unilaterally Draws the Line

16

CHAPTER 2

A Deal with Indonesia

36

CHAPTER 3

Australia vs. Portugal: Prelude to Invasion

57

CHAPTER 4

Death, Denial and Oil

90

CHAPTER 5
The Price of Recognition
117

CHAPTER 6
An Independent Adversary
155

Conclusion
186

A Note on the National Archives of Australia
193

Acknowledgments
198

Endnotes
199

INTRODUCTION

The Palácio do Governo, with its white arched colonnade running parallel to the Dili foreshore, is one of the few Portuguese colonial buildings to survive the violence of Timor-Leste's turbulent past. It is where former Victorian premier Steve Bracks and I met with legendary resistance leader and then prime minister Xanana Gusmão, in September 2007. Bracks was there in his new role as pro bono governance adviser to Gusmão. I was the adviser to the adviser. The Palácio do Governo is also home to the Cabinet room allegedly bugged by Australian spies three years earlier, when the fledgling state of Timor-Leste was trying to negotiate a maritime boundary with Australia.

Timor-Leste consists of the eastern half of the island of Timor, along with Atauro and Jaco islands and the enclave of Oecussi in West Timor. For over 400 years it was part of the Portuguese colonial empire and known as Portuguese Timor.

In 1974 Portugal granted its remaining colonies independence, and the following year the territory was

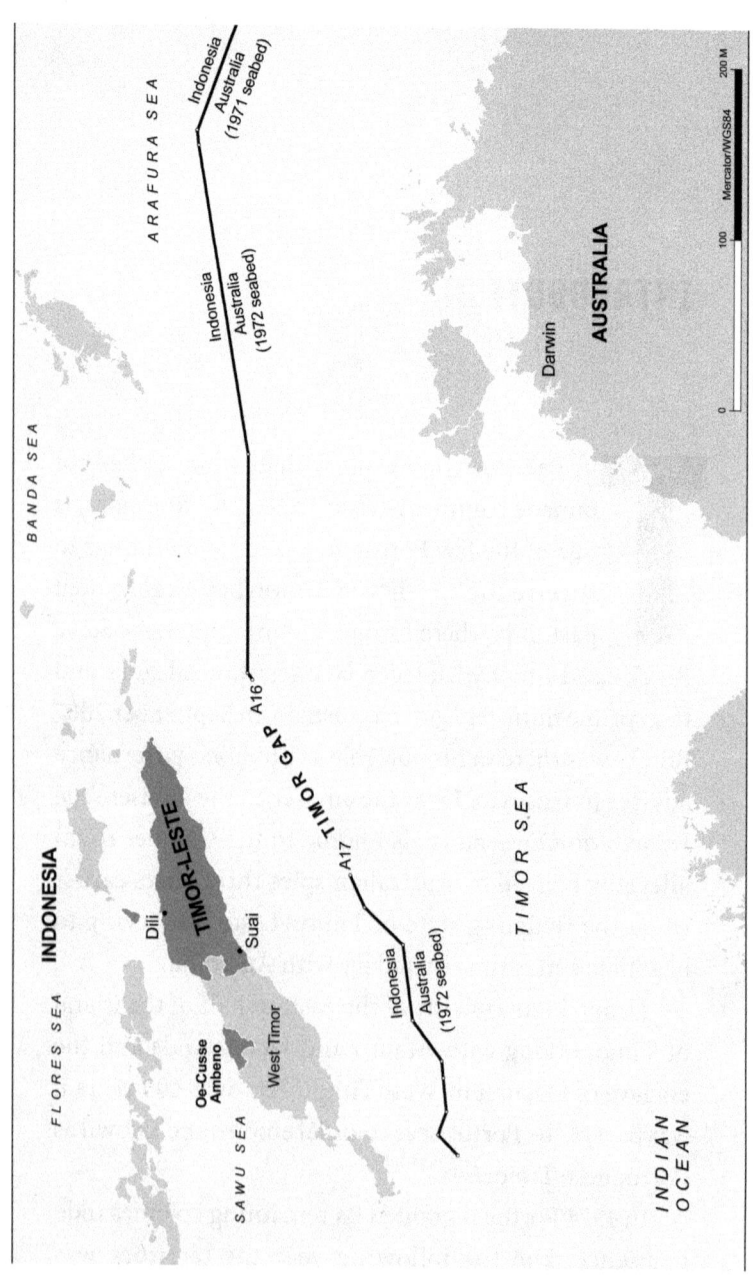

Figure 1: Timor Sea Treaty 1972

invaded and occupied by Indonesia. For twenty-four years the Timorese fought a war of resistance, culminating in a vote of self-determination in 1999 and admission to the United Nations as the Democratic Republic of Timor-Leste in 2002.

The capital, Dili, is just over an hour's flight from Darwin across the Timor Sea to Australia's north.

As Bracks and I had our briefing in the prime minister's office, I surreptitiously studied a map of the Timor Sea pinned to the wall. It showed the only permanent maritime boundary in the Timor Sea – a line agreed to by Australia and Indonesia in 1972, as shown in Figure 1. Portugal was excluded from the negotiations that left the 'Timor gap', and Australia's subsequent attempts to negotiate a boundary and close the gap have all failed. While 98 per cent of Australia's massive maritime boundary is settled, there is still no boundary line between Timor-Leste and Australia.

All oil and gas reserves shown on the map pinned to the wall were on the northern side of the line that runs halfway between the shores of Australia and Timor-Leste – the median line.

I was aware that much of this area had been unilaterally claimed by Australia when it issued exploration permits in the Timor Sea in the early 1960s. But I had no idea of the basis for Australia's claim under international law to oil and gas fields that were obviously much closer to Indonesia and Timor-Leste than to Australia. Nor did

I understand the Orwellian advice we received later that day: there was a ban on anyone in the Timorese government talking publicly about the need for a maritime boundary to close the Timor gap.

During that first visit, Prime Minister Gusmão hosted an official dinner for Bracks at his residence at Balibar, half an hour's drive from Dili along a treacherous winding road. We stopped on the way at a small village clinging to the side of the mountain so that his wife, Kirsty Sword Gusmão, could show us her plans for a new school and a modest museum at the site of a memorial commemorating the bond between Australian soldiers and their Timorese supporters during World War II. I recalled the powerful ads that ran on Australian television in the lead-up to Anzac Day in 2005. In the ads, Australian veterans who had served in Timor called on the Australian government to stop stealing Timor-Leste's oil and gas and explained that they owed their lives to the people of East Timor.

*

I made my forty-second visit to Timor-Leste in March 2017, nearly a decade after my first. The nation has been transformed. In 2008 Timor-Leste was in the top ten 'at risk' countries according to the Global Peace Index published by the Institute for Economics and Peace. In 2017 it was ranked alongside Singapore, Norway and the

United Kingdom as having a 'high state of peace'.

In the Boston Consulting Group's 2016 Sustainable Economic Development Assessment, Timor-Leste ranked seventh of 160 states for making the most progress in converting economic growth into wellbeing. A major study published in medical journal the *Lancet* rated Timor-Leste the most improved of 188 nations in the health-related Sustainable Development Goals index for the period 2000–2015. Timor-Leste came first in South East Asia, fifth in Asia and forty-third of all states assessed in the *Economist* Intelligence Unit's 2016 Democracy Index.

Talk of maritime boundaries was no longer taboo. Reflecting the significance of the issue, a new government agency, the Maritime Boundary Office, is tasked with securing permanent maritime boundaries with Australia and Indonesia.

The nation still faces massive challenges. Poverty levels remain some of the highest in the world. Stunting from decades of malnutrition is endemic. The education system struggles to cope with the high youth population and the complexity of running a school system in Portuguese, a language that not even all the teachers speak properly. The mountainous landscape and monsoonal climate make maintaining a road system difficult and expensive. Timor-Leste is the second most oil-dependent nation in the world, and reserves are running down.

Which is what makes the sea boundary dispute between Australia and Timor-Leste so critical – and why

I have spent much of the last decade researching the basis for Australia's claim to oil and gas resources on Timor-Leste's side of the median line in the Timor Sea.

I quickly found a wealth of information about the murder of five Australian-based journalists in 1975 at Balibo, and about the Australian troop deployment in East Timor following the independence vote in 1999. But there was little information, outside official Australian government publications and dense academic international maritime law or technical geography articles, about Australia's interest in oil and gas in the Timor Sea. I read and reread Paul Cleary's 2007 book *Shakedown: Australia's Grab for Timor Oil* on the tortuous negotiations between Australia and Timor-Leste immediately after independence, and wanted to know more.[1]

In a second-hand bookshop I chanced upon a copy of an intimidatingly large book, *Australia and the Indonesian Incorporation of Portuguese East Timor 1974–1976*.[2] It was an Australian Department of Foreign Affairs and Trade publication, which I was later able to download as a PDF running to 952 pages from the DFA website.[3] This work, which I refer to throughout as the Downer Compilation, includes 486 cables, briefs and other documents generated by officials in Canberra and in Australia's embassies in Indonesia, Portugal and other posts, that were approved for early release from the National Archives of Australia by foreign minister Alexander Downer in 2000. In the foreword, Downer said the publication was intended to 'go some way towards answering the many questions of those who have been concerned to obtain the truth about Australian official thinking and action in relation to the Indonesian incorporation of East Timor'.

The cables in the Downer Compilation confirm earlier claims by a range of former diplomats, journalists and academics that Australian officials were contemporaneously briefed on Indonesia's political and military strategies to subsume Portuguese Timor into the Indonesian republic. They also confirm that the Whitlam government secretly supported the Indonesian takeover of the territory. Australian prime minister Gough Whitlam and the ambassador to Indonesia, Richard Woolcott, are identified as the key protagonists driving Australia's foreign policy at this time.

The selected documents also show the divisions within the Whitlam government, at the ministerial and departmental level, that no doubt contributed to the large number of highly embarrassing documents, most authored by Woolcott, that were leaked to the Australian media. There are, however, only a few references in the compilation to the implications of the dramatic events to our north for Australia's multi-billion-dollar Timor Sea oil interests.

On my return to Melbourne I discovered the remarkable collection of documents in the Australian National Archives. It was clear that there were entire series of files dealing with the Timor Sea negotiations with Indonesia and Portugal, with the United Nations' Law of the Sea Conference, and with the companies holding permits, particularly one called Woodside. Hours clicking though digitalised DFA files led me to discover that in addition to ignoring files on Australia's diplomacy to secure oil and gas interests in the Timor Sea, the editors of the Downer Compilation had omitted information about Australia's Timor Sea oil interests from within documents that are included in the compilation.

An internal report on the implications for Australia of Portugal's decision in April 1974 to allow its colonies a vote of self-determination had an entire page of information about Australia's oil interests in the Timor Sea 'omitted'. Information relating to Portugal's claim to the median line in the Timor Sea was not included – a claim Portugal had backed up by issuing a permit to a US company, Oceanic Exploration, in January 1974.[4] This claim overlapped permits issued by Australia to seven Australian companies with some of the world's biggest international oil companies as partners. It was a claim that raised doubts as to the sovereignty over oil and gas resources worth potentially billions of dollars.

How was it that this dispute was not relevant or necessary to 'obtain the truth about Australian official thinking and action in relation to the Indonesian incorporation of East Timor'?

There are thousands of documents in the archives concerning Australia's interest in oil in the Timor Sea and historical relationship with Timor-Leste that I have had to leave out of this book. So I appreciate that the editors of the Downer Compilation could not include everything. But I can also see the elephant in the room.

Geopolitical issues and shifting alliances, Cold War brinksmanship and internal ALP and Coalition battles for foreign policy control are fascinating, and clearly worthy of inclusion in such a compilation. But so is Australia's direct interest in something far more tangible: oil and gas.

So perhaps it shouldn't come as a surprise that the Downer Compilation does not mention the fact that in June 1974 the Woodside/Burmah/Shell consortium successfully drilled and discovered oil and gas at Troubadour, part of the multi-billion-dollar Greater Sunrise field located on Timor-Leste's side of the median line.

No longer accepting the DFA's version of events, I decided to turn my interest into a PhD. That is still a work in progress. This book is based on my examination of thousands of documents in the National Archives of Australia about Australia's Timor Sea boundary negotiations with Indonesia and the dispute with Portugal. It is deliberately focused on one narrative: Australia's secret Timor Sea oil story. There are, of course, other factors at play in Australia's foreign relations with Indonesia, Portugal and Timor-Leste but, sadly for the Timorese, oil has been a dominant driver behind its neighbour's many betrayals since Australia unilaterally issued petroleum exploration permits in 1963.

As I dug deeper into the historical records I found briefs, reports, Cabinet submissions, correspondence and meeting minutes showing that many of Australia's politicians and senior diplomats in Jakarta, Lisbon and Canberra were very actively engaged in Australia's Timor Sea oil agenda throughout the period covered by the Downer Compilation (1974–1976).

Australia's keen interest in oil and gas resources in the Timor Sea was not only kept secret from the Australian public; Australia's major allies, the United Kingdom and the United States, were also kept in the dark. Of the 492 cables published in the Downer Compilation, 133 are marked 'AUSTEO' (Australian Eyes Only). An AUSTEO classification means that in no circumstances can material be supplied to foreign nationals. AUSTEO files are unusual, given the various agreements between Australia and its key allies to share intelligence information. But they are not unusual in the Downer Compilation, or in the hundreds of other records I have inspected.

The more files I looked at, the more it seemed that the records published in the Downer Compilation obscured, rather than revealed, the truth about Australia's official thinking and action.

Since long before the Trump administration coined the term 'alternative facts', the Australian government has been masterfully creating an 'alternative history' that excludes or at the very least downplays Australia's interest in oil-rich areas of the Timor Sea. This historical revisionism has resulted in the suggestion that Australia was primarily interested in oil and gas in the Timor Sea being dismissed as a left-wing conspiracy theory – even by progressive writers. For example, in his 2001 Quarterly Essay analysis of the records released by Downer, John Birmingham observes:

> The story of Timor was a tragedy with elements of farce, horror and villainy on all sides. It is not, however, amenable to simple conspiratorial explanations – such as the oft-repeated accusation that Australia was primarily motivated by a desire to cut Portugal out of negotiations for the Timor Gap oilfields. Such complicated, fearful and uncertain episodes are pregnant with the potential for myth-making.

Sometimes the simple explanation is the right explanation, and sometimes where there is smoke, there is fire.

My research in the archives reveals that since the 1960s Australian officials had access to seismic and other data from the oil companies that indicated there were significant oil and gas reserves in the area between Australia and Portuguese Timor; that Portugal was deliberately excluded from the seabed negotiations between Australia and Indonesia in 1972; that Australia was involved in an escalating diplomatic dispute with Portugal over rights to the most prospective areas of the Timor Sea; and that

Australian officials believed that if Indonesia took over Portuguese Timor, Indonesia would agree to close the Timor gap with a straight line, putting all the oil and gas resources on Australia's side of the line.

It is therefore not conspiratorial or 'myth-making' to argue that Australia's response to Indonesia's invasion of Portuguese Timor was influenced by its interest in the Timor gap oilfields. Australia had a multi-billion-dollar interest in Indonesia taking over Portuguese Timor.

Shadow Foreign Affairs Minister Laurie Brereton criticised the DFA for focusing on the period before the Indonesian invasion, when the ALP was in power, and largely ignoring the years immediately afterwards when the Liberal–National Coalition government was in power. In the three years following Indonesia's invasion of East Timor, an estimated 200,000 Timorese died. There were 'systematic and widespread unlawful killings and enforced disappearances of surrendered civilians and combatants' by Indonesian troops and, in 1978 and 1979, deliberately induced famine.[5]

These years coincided with repeated attempts by the Fraser government to recognise Indonesian sovereignty, a necessary step in order to negotiate to close the Timor gap. Throughout this period the Australian government pursued a ruthless, single-minded policy agenda to protect the integrity of the exploration permits unilaterally issued in 1963, regardless of any moral consequences. This was made explicit by Australia's ambassador to Indonesia,

Richard Woolcott, in a cable leaked to the Australia media in 1976:

> Government is confronted by a choice between a moral stance, based on condemnation of Indonesia for the invasion of East Timor and on the assertion of the inalienable right of the people of East Timor to self-determination, on the one hand, and a pragmatic and realistic acceptance of the longer term inevitabilities of the situation, on the other hand. It's a choice between what's described as Wilsonian idealism or Kissingerian realism. The former is more proper and principled but the longer term national interest may well be served by the latter.[6]

The archives reveal that the Fraser government schemed to keep evidence of the brutality of the invasion out of the Australian media and off the agenda at the United Nations because it was politically difficult to commence maritime boundary negotiations when there was so much domestic opposition to Australia officially recognising Indonesia's sovereignty over East Timor. Negotiations would, of course, signal Australia's recognition of Indonesia's sovereignty – something the Australian public, and many in the government, opposed.

Australia's efforts to secure oil-rich areas of the Timor Sea north of the median line have been even more brazen since the Timorese voted to become independent in

1999. The Timor Gap Treaty ceased to exist following Indonesia's withdrawal from the territory, and under international law it was clear by then that the new nation would have rights to an Exclusive Economic Zone (EEZ) including petroleum resources up to the median line.

Yet Australia continued to collect revenue from oil and gas fields on the Timorese side of the median line. Since 2005, Australia has collected $1.4 billion from oil and gas fields from a 'resource sharing' zone Australia negotiated with the United Nations before Timor-Leste became independent. On top of this, Australia has collected 100 per cent of the revenue from oil and gas fields in other areas Timor-Leste claims are within the area it is entitled to, according to the international law of the sea. Australia has also reaped $25 billion in downstream benefits from ConocoPhillips' LNG plant in Darwin.

And Australia is still claiming the right to revenue from the as yet untapped Greater Sunrise field, which is wholly on the Timorese side of the median line.

Drawn-out treaty negotiations directly involving Australia's foreign minister, an effective grassroots campaign calling for a treaty based on the median line, sensational allegations of spying, the seizure by Australia of Timor-Leste's documents concerning the spying allegations, a successful hearing in the International Court of Justice, and a long overdue policy commitment at the ALP's July 2015 National Conference to begin maritime boundary negotiations with Timor-Leste and to resubmit to the jurisdiction of the international umpire have kept Australia's Timor Sea oil agenda in the spotlight.

Yet the Timor gap remains.

In January 2017 the Australian government gave a public commitment to negotiate a permanent maritime boundary with Timor-Leste. The negotiations are occurring under the auspices of a Compulsory Conciliation Commission convened under the UN Convention on the Law of the Sea. Gusmão is the chief negotiator and Minister Agio Pereira is Timor-Leste's agent. Australia's participation in the conciliation process is compulsory, but neither party is bound by the outcome.

Australia's ruthless pursuit of oil and gas resources in the Timor Sea over the last half-century suggests that it will not be easy for Australia to agree to close the Timor gap in accordance with international law.

CHAPTER 1
AUSTRALIA UNILATERALLY DRAWS THE LINE

Until 1978 the Northern Territory was administered by the federal government, which is why the first offshore petroleum exploration permits Australia issued in the Timor Sea were stored in the National Archives in Darwin. There I found records that led me to a chapter of Australia's Timor Sea oil history involving the unlikely combination of oysters, oil and the brilliant geological analysis of a Russian aristocrat.

In an archive box in Darwin I found the original application for a petroleum exploration permit in the Timor Sea from Woodside (Lakes Entrance) Oil and its fully owned subsidiary Mid-Eastern Oil, dated 8 November 1962. There was a hand-drawn map showing rectangular blocks and a squiggly yellow line following the 100-fathom (approximately 200 metres) depth line. I was amazed to see that one of the stand-alone blocks targeted for exploration included the multi-billion-dollar Greater Sunrise oil and gas field – over a decade before it was discovered.

How had Woodside been able to pinpoint one of the biggest oil and gas fields in the Timor Sea in 1962, before any seismic tests had been conducted? And on what basis was the Australian government able to grant exploration permits to blocks that were much closer to Portuguese Timor and Indonesia than Australia?

The short answer to the first question is Nicholas Boutakoff.

Boutakoff, the geologist who prepared Woodside's 1962 application, was born in Washington DC in 1903, the son of a Russian diplomat. His father and brother were executed by the Bolsheviks in the 1917 Russian Revolution and Boutakoff's Russian nationality was withdrawn by the Soviets. He was stateless until he became a British citizen in 1943. In the meantime, he studied geology in Europe, worked for mining exploration companies in the Congo, was involved in the Spanish Civil War and later worked for Royal Dutch Shell in Trinidad. He joined the Victorian Geological Survey department in Melbourne in the 1940s in search of a quieter life.

But he had the spirit of an explorer. Following the discovery of Australia's first free-flowing petroleum in 1953, Boutakoff was sent to Western Australia to investigate oil exploration methods and techniques that could be applied in Victoria. As historian Robert Murray explains,

'Disturbance of the earth's surface is one of the first things explorers look for as it may indicate a corresponding faulting or folding at great depths, which could form the traps required to hold petroleum.'[1] Boutakoff later told colleagues he stood on Rough Range and could visualise the onshore geological contours of the land stretching out to sea. He returned from the west so confident there was oil under the Timor Sea and off the north coast of Western Australia that in March 1955 he formed a company, Northern Holdings Pty Ltd, and applied for exploration permits from the federal government in Darwin.[2] Boutakoff's papers are preserved in nine boxes in the State Library of Victoria. Among them I found a letter to his US partner in March 1955, in which he explains that the concession he was applying for from the federal government was 'only 30 to 60 miles from the southern coasts of Timor'. I couldn't find any record of this application in the Darwin or Canberra archives. The financial backing Boutakoff expected from his US partner fell through and by 1958 Northern Holdings had been wound up.

OYSTERS AND THE LAW OF THE SEA

As it turns out, Boutakoff's confidence in Australia's right to issue petroleum exploration permits 30 to 60 miles from the coast of Timor was based on legislation originally intended to assert Australia's rights to pearl oyster beds against competing Japanese interests.

By the mid-twentieth century, the international law of the sea recognised three main zones: internal waters lying landward of the baseline of the territorial sea, over which a nation exercised full sovereignty; the territorial sea, generally between three and twelve miles from the low-water mark, over which a nation exercised full sovereignty (subject to a right of innocent passage for shipping); and the high seas, open to all nations in times of peace and not subject to national sovereignty.

Technological advances during World War II had made it possible to drill for oil under the seabed in the vast oceans of the world, and demand for petroleum was escalating. US president Harry Truman issued a proclamation in 1945 that asserted US sovereignty over 'natural resources' on its surrounding continental shelf. Other nations with wide continental shelves embraced the concept and quickly made similar claims. From its beginnings in the Truman Proclamation, the concept of the continental shelf was plagued by uncertainty over the question of its limits. It fell upon the International Law Commission that was codifying the international law of the sea to determine the limits of the continental shelf and the definition of 'natural resources'.

While the Truman Proclamation was clearly about US access to oil, the natural resources Australia was primarily interested in at that stage were oysters. Australia and Japan had been in dispute over the right to harvest pearl oysters off Australia's northern coast for decades. Australia's solicitor-general Kenneth Bailey saw an opportunity to allow Australia to claim sovereignty over disputed pearl fisheries off the western and northern coasts of Australia by extending the definition of 'natural resources' covered by a continental shelf proclamation to include oysters – described by Australia's lawyers as 'sedentary fisheries'.[3]

In September 1953, Australia issued its own continental shelf proclamation based on Truman's. The *Pearl Fisheries Act (No. 2) 1953* set out that Australia's continental shelf extended to the 100-fathom depth contour (approximately 200 metres). The proclamation asserted one of the largest maritime areas in the world: more than 15 million square kilometres, almost double the size of the Australian landmass.

Australia then successfully lobbied at the International Law Commission to have 'sedentary fisheries' (a.k.a. oysters) included in the definition of natural resources in the Convention on the Continental Shelf 1958. This convention provides that coastal states have exclusive rights to natural resources on the continental shelf, which was given a two-pronged definition. The first part of the definition was consistent with Australia's *Pearl Fisheries Act*, providing that the continental shelf extended from the coast to a depth of 100 fathoms.

The second part was far more nebulous and led to years of confusion and controversy. It provided that the continental shelf could extend 'beyond that limit' (i.e. 100 fathoms) 'to depths that allowed the exploitation of natural resources'. As technology improved and the depths at which drilling could occur increased, the area meeting the definition of continental shelf stretched out well beyond the 100-fathom line.

It is important to note that the convention only applied to the seabed. The International Law Commission still had to settle the rules applying to water column (or fisheries) boundaries.

WOODSIDE SEEKS A STAKE IN THE TIMOR SEA

This was the state of play in 1962 when Boutakoff quit the Victorian bureaucracy and accepted an offer to join Woodside, taking his maps and his enthusiasm for the Timor Sea and Australia's north-west with him.

Woodside had been in operation since 1954 but was yet to strike oil. The company was formed by accountant Rees Withers, who was later described by a Victorian Supreme Court justice in a pro-Woodside judgment as a 'single-minded opportunist who was not handicapped by any scruples or feelings of loyalty'.[4] The name Woodside heralded from the company's first drill site, an isolated town on the windswept southern coast of Victoria. Melbourne stockbroker Geoff Donaldson, who underwrote Woodside's initial float, became chair in 1956. Donaldson described the early operations of the company as a 'circus', and by 1959 Woodside was broke.[5] But an oil discovery at Moonie in Queensland in December 1960 sent oil share prices soaring, enabling Woodside to regroup and survive.

Boutakoff briefed the Woodside board in September 1962 on the exploration applications he had prepared for Northern Holdings in the mid-1950s. Donaldson recalled, 'We couldn't possibly afford it, but for a few hundred pounds, which was all we had, we jumped in.'[6]

It was Boutakoff who drafted the letter of application dated 8 November 1962 I found in the archives in Darwin that confidently stated the areas applied for that were much closer to Timor than Australia were 'part of The Australian Continental Shelf'.

Kenneth Bailey, the architect of Australia's innovative oyster claim, initially advised the Department of the Interior that Australia had the power to issue the Timor Sea permits. Bailey's advice was 'confirmed' by attorney-general Garfield Barwick, 'who expressed the view that, from the international standpoint, the areas were part of the continental shelf of Australia'.[7] This view was to prove highly speculative.

Woodside's new venture, which became known as the North West Shelf Project, was announced in a paper delivered by Boutakoff at the annual conference of the Australian Petroleum Exploration Association in Brisbane, in March 1963. The conference was the Australian oil industry's annual get-together and attracted the big US oil companies as well as representatives from Shell and Britain's Burmah Oil. Woodside needed partners, and the conference was the perfect pitch opportunity. Boutakoff told the conference that his mapping of the shelf floor revealed the existence of ridges on the continental shelf that 'may prove to be suitable for considerable accumulation of petroleum'.[8] Despite the understated language, Boutakoff was convincing, and by the end of the year Woodside – a nearly insolvent company without a run on the board – was joined in its venture by oil giants Shell and Burmah Oil, sending a powerful signal to the Australian government that the Timor Sea was potentially a very valuable asset.

The year Boutakoff delivered his paper, 1963, was also the year the Australian Cabinet secretly formed the view that it was inevitable that Portuguese Timor would become part of Indonesia. The former secretary to the parliament's committee on foreign affairs and defence, Robert King, stressed this point in submissions to various parliamentary inquiries into Australia's relationship with East Timor.[9] But until I read the DFA files in the archives I was unaware that the Cabinet decision was in part prompted by a request from President John F. Kennedy for Australia to take the lead in pressuring Portugal to allow a vote of self-determination in Portuguese Timor.

THE ROAD NOT TAKEN

When Kennedy came to power in 1961 the massive archipelagic nation of Indonesia, which had been formed in

1949 from the former Dutch East Indies, was the focus of Cold War tension in South East Asia. Russia was selling arms to Indonesia, and the Communist Party of Indonesia was growing in influence. The Eisenhower administration had responded by approving nefarious CIA operations designed to bring about 'regime change' in Indonesia, including support for a failed rebellion against President Sukarno in 1958.

Kennedy adopted a radically different approach. He encouraged ties with the Indonesian army to counter the influence of the Soviet Union and China. He cancelled the CIA operation and engaged with Sukarno, offering a ten-year aid program of education and welfare, and brokered a deal to get the Dutch to give up Dutch New Guinea, the one part of the former Dutch East Indies that the Netherlands had not relinquished in 1949.

Kennedy was also a strong advocate for decolonisation. DFA files reveal that Kennedy gave veteran diplomat Averell Harriman the job of getting Australia to agree to take the lead on pushing Portuguese president Antonio Salazar to implement a similar ten-year aid program of education and welfare in Portuguese Timor, followed by a vote of self-determination.

The United Nations had moved against President Salazar in 1960; the Declaration on the Granting of Independence to Colonial Countries and Peoples specifically listed Portuguese Timor and seven other Portuguese-controlled territories as Non-Self-Governing Territories that had the right to self-determination. Salazar ignored the declaration, arguing Portugal's colonies were overseas provinces and therefore not subject to it. He continued Portugal's military efforts to crush independence movements in its African colonies.

Kennedy needed Australia to step up and take the lead with Salazar. The United Kingdom and the United States had to be muted in their criticism of Portugal's African wars and defiance of the UN because both had long-standing treaties with Portugal granting military access to the strategically placed Azores Islands in the Atlantic that they did not want to put at risk.

Menzies added foreign affairs to Barwick's responsibilities in December 1961. With his attorney-general hat on, Barwick was advising it was legal for Australia to issue petroleum exploration permits in potentially oil-rich areas of the Timor Sea just off the coast of Portuguese Timor. With his external affairs minister hat on, he was advising it was inevitable that Portuguese Timor would become part of Indonesia, and that Kennedy's proposal was therefore a waste of time.

Barwick's coolness towards the Kennedy proposal was obvious to the Americans. Harriman invited Australian journalist Denis Warner, the special correspondent for the Melbourne *Herald* group, to lunch in Washington and outlined Australia's key role in Kennedy's 'preventative diplomacy' plan for Portuguese Timor. Harriman complained that the United States had been 'trying to encourage such action for a long time, but there has been no receptive Australian response so far'.

Warner's story was published in Sydney's *Daily Telegraph* on 4 February 1963 under the headline 'Australia has big role in Portuguese Timor'. Warner reported that the Kennedy administration believed the only way to thwart an Indonesian attempt to take over Portuguese Timor was for Australia to urge Portugal to implement urgent short- and long-term plans to raise living standards. Quoting Washington sources, Warner encouraged Australia to take a more active role in Asia, arguing that 'we should approach the Portuguese government at the highest level to insist it is time the people of Portuguese Timor were brought into the 20th Century'.

Barwick was furious. He cabled Australia's ambassador in Washington, Howard Beale, on 5 February instructing him to tell Harriman that 'we have read Warner's article and are aware that he is the source of it' and that 'we do not relish receiving the views of the United States Government on the future of Timor' through the media. On the substance of the proposal, Barwick instructed Beale to tell the state department:

> I completely fail to see how pouring money into Timor to raise the standard of living is going to solve this problem. Does the State Department imagine that there is any future for an independent Timor with a population of only 540,000 ... have they any

reason to suppose that Portugal would allow the United States to start development of the territory on what seemed the right lines to Washington? My own impression is that Portugal would regard this as unwarrantable interference and also that the fabric of the Portuguese Timor administration is incapable of adjustment to such a developmental programme.[10]

At talks between the United States, the United Kingdom, New Zealand and Australia in Washington on 12 February 1963, there was general agreement that it was inevitable Portugal would have to relinquish control of Portuguese Timor. No nation was prepared to commit forces to prevent an Indonesian invasion. According to the Australian record of the discussion:

Various ideas were put forward, ranging from the suggestion, at one extreme, that the West should take early action to raise Timor in the U.N. and keep the issue alive there, to the suggestion at the opposite extreme that it would be in Western interests if Indonesia were to devour the territory quickly and with a minimum of publicity.[11]

There is no suggestion in the Australian records of this meeting that the Australian representatives shared with their diplomatic allies that just a week earlier, on 5 February 1963, the Australian Cabinet had accepted advice from Barwick 'that in the current state of world opinion, no practicable alternative to eventual Indonesian sovereignty over Portuguese Timor presented itself'.[12] Cabinet's caveat that 'it would not be acceptable to Australia or the West for Indonesia to proceed against Portuguese Timor with arms' appears tokenistic at best, given Barwick's advice, less than two weeks later, that Indonesia intended to take over Portuguese Timor 'at some stage' in what 'would be an easy military operation'.[13]

Following the Washington talks, Barwick told Cabinet he was convinced 'Harriman's idea has no future in it.' Had Australia taken up Kennedy's challenge and succeeded in persuading Salazar it was in Portugal's interests to allow a vote of self-determination in Portuguese Timor, the Timorese would have voted in a UN-supervised plebiscite in 1973. But this scenario would have threatened Australia's nascent interest in the potential oil wealth of the Timor Sea. Politically it would be more palatable for Australia to assert rights to oil and gas resources 30 miles off the coast of a fading, fascist, unpopular colonial power than to deny those resources to a potentially independent, recently liberated, desperately poor neighbour.

AUSTRALIA'S MASSIVE MARITIME BOUNDARY CLAIM

Barwick next had to deal with doubts about Australia's right to issue exclusive exploration permits on its continental shelf. International law was still unclear, and

Barwick acknowledged in September 1963 that it was

> no secret that what has been retarding off-shore oil exploration has been in part the legal advice that some foreign companies have received from their own advisers that on the continental shelf a State is not legally able to make exploration permits effective by denying to foreign rivals access to the areas concerned.[14]

Of even more concern were moves by an increasing number of nations to claim an Exclusive Economic Zone (EEZ) over a fixed 200 nautical mile *distance* from the coast, in preference to a continental shelf claim based on water depth. Under the EEZ regime it was proposed that in the case of overlapping EEZ claims, as in the Timor Sea, which is less than 400 nautical miles wide, the boundary would be the median line – halfway between each coast.

Complicating matters further, there was also uncertainly under Australia's constitution about which level of government, federal or state, had the constitutional right to issue offshore permits beyond the territorial sea. This prompted Woodside to write to the federal government in December 1963 about the implications for their permit of a federal legislative takeover of offshore resources.

AUSTRALIA REJECTS THE MEDIAN LINE

A compromise was reached at a Premiers' Conference in

June 1965 where it was agreed to introduce joint federal–state legislation to regulate offshore petroleum exploration. As a first step, the federal government had to decide on the boundaries of the proposed scheme. The only area around Australia's massive coastline where there was a significant risk of competing claims was in the north, in the Timor Sea. Debate centred on whether the boundary should at a minimum include all existing permits areas in the Timor Sea, or if Australia should pull back to the median line between the coasts of Indonesia and Portuguese Timor and Australia.

By 1965 most of the Timor Sea between Australia and Timor was a patchwork of permits. Woodside and its partners, Shell and Burmah Oil, had been joined by the Atlantic Richfield Company of America (ARCO), the Société Nationale des Pétroles d'Aquitaine (Aquitaine) and an Australian company, Timor Oil. All of these companies were very actively searching for petroleum, conducting marine seismic surveys and aeromagnetic surveys of the region.

A condition of the generous subsidies under the *Petroleum Search Subsidy Act 1958* was that survey results had to be sent to the Australian Bureau of Mineral Resources, Geology and Geophysics in Canberra. The bureau also conducted its own survey in 1965, with encouraging results. Robert Murray concludes that the early seismic results reported during 1964 and 1965 'pointed to an oilman's dream'.

The minister for national development, David Fairbairn, recommended that the proposed joint state–federal petroleum regime should extend no further than the median line, regardless of the potential riches in areas further north. He argued that the permit holders were primarily interested in 'security of tenure' and that it was not in Australia's national interest to risk antagonising Indonesia by pushing beyond the median line. Fairbairn was also concerned that if Australia was later unable to issue a production licence the government could be liable to compensate companies for their losses.

History, we now know, was on Fairbairn's side. But his views were not shared by the DFA. The DFA opposed negotiations with Portugal, arguing they would 'imply a degree of acceptance on our part of Portugal's right to share in decisions permanently affecting the future of the area'. This would run counter to Cabinet's February 1963 assessment that 'no alternative to eventual Indonesian sovereignty over Portuguese Timor presented itself'.

The archival records reveal that the DFA was also unconcerned about the consequences for Australia's relationship with Indonesia. This is in marked contrast to the DFA's advice in relation to Indonesia's invasion of East Timor in 1975, which consistently prioritised good relations with Indonesia above any other consideration, including the consequences for the Timorese. In 1965, the DFA explicitly advised against opening negotiations on a Timor Sea boundary with Indonesia because Indonesia was 'unlikely to envisage any solution as acceptable other than that of the median line'.[15]

One explanation for the DFA's lack of concern about Indonesia's reaction to Australia's audacious Timor Sea claim can be found in the dramatic events unfolding in Indonesia at the time. An attempted coup on 30 September 1965 led to the regime change in Indonesia that the CIA had been supporting before President Kennedy's shift in policy. The coup resulted in Sukarno being replaced as president by General Suharto. Suharto unleashed a violent 'anti-communist' campaign with the clandestine support of the United States and its Western allies, during which up to a million communist party supporters, leftist intellectuals, trade union officials and ethnic Chinese were killed in what the CIA itself describes as 'one of the worst mass murders of the twentieth century'.

While the political crisis in Indonesia is not mentioned in the DFA's briefs concerning the Timor Sea, it would have bolstered the advice not to engage with Indonesia at that time. Australia welcomed the demise of Sukarno and set about building a positive relationship with the new, vehemently anti-communist leader, Suharto.

Fairbairn lost the debate about where to set the boundary for Australia's petroleum exploration in the Timor Sea. The lure of oil, the advice that Australia had rights to issue permits on its continental shelf, coupled with a general disdain for the views of Indonesia and Portugal, led Cabinet to lock in behind the DFA and agree to 'confirm existing permits and claim jurisdiction over areas covered by these permits without prejudice to such rights as Australia may possess over areas beyond'.

Less than two weeks later, Australia's unilateral action was questioned in an article in the *Australian* headlined 'Australia risks oil challenge by Indonesia'.[16] The article claimed the continental shelf boundary 'could put some of our oil exploration leases only a short distance from Timor' and that 'Indonesia would still have the formal right to challenge' the boundary. This was a clear indication to the Timor Sea permit holders that despite assurances from the highest levels of government, their permits were not secure.

AUSTRALIA MAKES ITS CHOICE

Robin Ashwin, one of the members of the advisory board for the Downer Compilation, is a South Australian Rhodes Scholar and former diplomat. In a 2006 interview Ashwin recalls that in 1967, shortly after being appointed to head the Defence Policy Planning section of the DFA, he attended an interdepartmental meeting about the international law of the sea and, 'more importantly perhaps, on the allocation of petroleum exploration licences in the Timor Sea'.[17] The other departmental representatives were taking what Ashwin described as the 'maximalist' line in relation to the Timor Sea permits – 'that Australia should claim everything it could', pushing the boundary out to within 50 kilometres of the Timor coast. According to Ashwin, the maximalist position was all about being able to 'control oil exploration'. Ashwin argued that 'you should look to what would be affecting your country not this month or this year or next year but 10 years or 20 years down the road'. He said it was in Australia's national interest to maintain a long-term good relationship with Indonesia and Portuguese Timor and that required 'drawing a median line in the middle of the sea between us and them.' As he recalls, he was 'laughed out of court'.

In late October 1967 Fairbairn introduced to parliament seven bills that were referred to as the Australian Offshore Petroleum Settlement. In his second reading speech, Fairbairn stated that:

> In 1965 I expressed the hope that further exploration would bring fresh discoveries of both oil and natural gas. This hope has been well justified. There have been further discoveries of petroleum; and it is especially gratifying that the companies engaged in offshore operations have demonstrated their confidence in the successful outcome of these inter-governmental negotiations by engaging in a steadily expanding programme of offshore exploration ...

By then the Australian government had access to the results of seismic surveys north of the median line that gave reason to be optimistic about the petroleum potential of the area.

In 1965, Fairbairn had been concerned about the implications for Australia's relationship with Indonesia if Australia confirmed permits north of the median line in the Timor Sea. However, neither Indonesia nor the Timor Sea rated a mention in his 1967 speech. Fairbairn tabled a map that showed the areas where permits had already been issued. In the Timor Sea, the coordinates coincided with the northern-most limits of the permits issued by Western Australia and the federal government. In other words, Nicholas Boutakoff's hunch about the presence of oil determined the boundary, rather than legal principles or geography.

The bills were passed. But the scheme's capacity to provide security of tenure was short-lived. In 1969 Indonesia and Portugal both issued continental shelf proclamations of their own. In December 1969 the United Nations passed a Moratorium Resolution requiring nations to refrain from exploration of the resources of the 'sea bed and ocean floor, and the subsoil thereof, beyond the limits of national jurisdiction'.

By the end of the decade, moves at the United Nations to codify the international law of the sea and the increasing risk of Indonesia or Portugal issuing permits that overlapped Australia's permits were to force Australia to the negotiating table.

CHAPTER 2
A DEAL WITH INDONESIA

I inspected hundreds of files in the reverential silence of the art deco reading room in the National Archives in Canberra. Thirty to forty files would arrive on a trolley in no particular order. Later, I learned to pause and sort them into series and issues, but initially I couldn't resist. I'd pick up a file, turn to the back page so I could follow the story chronologically, and be unable to put it down.

Having worked for five years in the late 1980s in an agency in the federal attorney-general's department, I was familiar with the filing system from the pre-digital world, when the business of government was conducted on paper. Departments opened a file on a subject as it became topical, and originals and copies of documents, ranging from draft Cabinet submissions to the seating arrangements for official lunches, were added on top. At around three centimetres deep a new folder for the file would be opened.

The files on Australia's Timor Sea oil agenda form the backbone of this book. Lots of material is still intriguingly classified, with paragraphs or sometimes entire pages blacked out. Occasionally I came across a declassified document that seemed to be the same as a document on another file that had sections redacted. Sometimes I found a document, the significance of which escaped me until I saw it again on a different file, in a different context.

Despite the redactions, it was clear that by the late 1960s Woodside and the other permit holders were regularly seeking exemptions from permit obligations to conduct exploration activity within a prescribed time, until the boundary issue resolved. The DFA was also concerned about the risk of overlapping permits unless a seabed boundary could be agreed. A US oil company had sought permission from Portugal to explore for petroleum in the Timor Sea between Portuguese Timor and the median line, and Indonesia was also starting to issue exploration permits in the area.

Australia's prime minister by then was John Gorton, who controversially pushed for the federal government to assume greater responsibility over offshore oil and gas. Gorton's foreign minister, William McMahon, was not in Gorton's centralist camp. In fact, he was not in Gorton's camp at all. Veteran political journalist Laurie Oakes describes McMahon as a 'devious, nasty, dishonest ... individual' who 'lied all the time'.[1] McMahon actively pursued Australia's Timor Sea agenda, both as foreign minister and, from March 1971, as prime minister, after a no-confidence vote led to Gorton's resignation.

There was a lot at stake. In November 1971, the Department of National Development estimated that recent discoveries on Australia's 'North-West Shelf' – which included the Timor Sea – 'may double the reserves of gas and we have hopes of finding oil in this area'.[2]

The archival records reveal the enormous diplomatic effort Australia expended to secure oil-rich areas of the Timor Sea. They also reveal a pattern of duplicitous diplomacy that continues today.

AUSTRALIA BURIES EXPERT ADVICE

Australia and Indonesia held exploratory seabed boundary talks between 19 and 24 March 1970. Australia's negotiating position was set out in a McMahon Cabinet submission dated 27 February.

Article 6 of the Convention on the Continental Shelf provides that a median line boundary should apply where the 'same continental shelf' was 'adjacent to the territories of two or more States whose coasts are opposite each other'. The DFA was aware that Indonesia would be seeking a median line boundary in accordance with the convention.

Australia sought to avoid a median line boundary by arguing that it did not share the 'same continental shelf' with Indonesia. Australia claimed the Timor Trough, a deep channel running parallel to the Timor coast about 30 to 50 nautical miles offshore, marked the edge of Australia's continental shelf. The existence of the Timor

Trough, according to McMahon, meant 'the question of fixing a common boundary in the Timor Sea does not really arise'.

The question of whether the Timor Trough marked the edge of Australia's continental shelf was by no means clear and was debated by geologists at the time. The contrary view was that the trough was just a dint in Australia's continental shelf, which in fact extended beneath the island of Timor.

If Australia's continental shelf did extend beneath Timor, then Article 6 of the convention would apply. Australia would be required to negotiate a median line boundary with Indonesia, and the permits issued north of the median line would be invalid.

Advice from Australia's own expert body indicating the Timor Trough did not mark the edge of Australia's continental shelf would have been explosive.

Sandwiched between the draft and final versions of McMahon's submission on a DFA file on 'continental shelf boundary negotiations' is a Bureau of Mineral Resources report titled *The Timor Trough – A Summary of Current Geological Knowledge*.[3] The paper, dated February 1970, concluded that there was 'no evidence of oceanic crust in the floor of the Timor Trough and hence no evidence to suggest there is any continental margin between the Sahul Shelf and Timor'.

If, as the paper seemed to be saying, the Timor Trough did not mark the edge of Australia's continental shelf, Australia had no basis for its claim to sovereignty beyond the median line in the Timor Sea. Australia's core argument at the talks about to commence with Indonesia would be undermined.

Professor Mike Sandiford and Dr Brendan Duffy, geologists from the University of Melbourne, confirm that the Bureau of Mineral Resources was telling Australia's negotiators that there was no strong geological reason to refute Indonesia's claim of a common continental shelf between Timor and Australia. They say the weight of geological opinion, then and today, describes the Timor Trough as a buckling of Australia's continental shelf, which extends beneath the island of Timor. So expert geological opinion backs the position put by Indonesia.[4]

But the bureau's report is not discussed in McMahon's Cabinet submission.

I found another copy of the bureau's report on a file dealing with Australia's negotiations with Portugal in 1974.[5] There was a DFA covering note dated 5 March 1970 addressed to Laurence McIntyre and Kenneth Bailey, who were leading Australia's Timor Sea boundary negotiations with Indonesia. It simply stated: 'Please find attached a paper prepared by the Bureau of the Mineral Resources at your request on the geology of the Timor Trough ... This paper was originally sent to the Department on 18 February but was evidently not received.'

This covering note has ticks next to McIntyre's and Bailey's names, indicating that this time, at least, it was received by them. While the report may not have been received in time for inclusion in McMahon's submission, it was certainly received before the start of the negotiations with Indonesia on 19 March. Yet in his opening address at the explanatory talks with Indonesia, McIntyre completely ignored this advice from Australia's expert geological body, and instead mounted the argument that the Timor Trough marked the edge of Australia's continental shelf, and should therefore be the boundary.[6]

And in Australia's public opening address at the UN Australia Timor-Leste Compulsory Conciliation in August 2016, Australia was still claiming that the 'physical continental shelves of Australia to the south and Timor-Leste and Indonesia to the north ... are separated by the Timor Trough'.[7]

Australia's claim to oil and gas fields north of the median line is built on a chimera. On a false premise that – remarkably – still underpins Australia's claim beyond the median line today.

PORTUGAL EXCLUDED FROM TIMOR SEA BOUNDARY TALKS

The first round of exploratory talks with Indonesia in March 1970 was abandoned without resolution of the continental shelf issue. The next round of talks was scheduled for February 1971.

When Portugal's ambassador to Australia, Carlos Wemens, heard about the March 1970 exploratory talks he sought to join the negotiations or negotiate bilaterally with Australia. He was politely rebuffed by the DFA.

The records show that Australia rejected numerous approaches for Timor Sea boundary talks from Portugal, including a request to the Australian embassy in Lisbon in August 1970 for Portuguese officials to travel to Canberra for 'friendly discussions'.[8] Lisbon cabled Canberra, noting that, 'surprising as it may seem', the Portuguese took 'the view that there is one shelf there and not two, and that the Timor trough does not constitute a division between one shelf and another'. The embassy also advised that Portugal had been approached by a US oil company for permission to conduct exploratory drilling in an area south of the Timor Trough. The company, Oceanic Exploration, was owned by Denver-based Jack Grynberg, a self-made millionaire who secured his fortune in 1962 at the age of thirty after reworking an abandoned gas field.

Portugal responded to Australia's stonewalling by publishing in the Portuguese government's official gazette three permit applications by Oceanic for petroleum exploration. This created a dilemma for the DFA. If Australia lodged an official protest about the Oceanic applications, it would be difficult to avoid being dragged into maritime boundary negotiations with Portugal, and McMahon's strategy was to settle a boundary with Indonesia first. As Oceanic was an American company it also meant that

Australia's interest in the Timor Sea potentially clashed with that of its key ally, the United States. This perhaps explains why so many documents concerning Australia's Timor Sea oil agenda are marked AUSTEO.

Portugal escalated the dispute in early November 1970, sending an official note to the Australian embassy in Lisbon requesting negotiations about the matters in 'dispute' and, according to a DFA brief, 'formally reserving Portugal's position with respect to what it regarded as "the unilateral appropriation by Australia of areas over which the island of Timor has rights"'. Portugal then said:

> It seems therefore that there is a difference of views between Portugal and Australia regarding the criterion applicable to the division of the continental platform between Timor and Australia.
>
> The Portuguese Government intending to grant concessions in the above-mentioned areas, the competent departments deem it to be most appropriate that consultations should take place as a matter of urgency on this subject with the Australian Government in order to reach … an understanding between both parties. Such conversations … should take place before the end of the present year, preferably during the month of December next. The Ministry of Foreign Affairs would accordingly be grateful if the Embassy of Australia would offer its observations on the matter in dispute.[9]

If the note was intended to force Australia to the negotiating table, it failed. It is not mentioned in McMahon's November 1970 Cabinet submission and appears to have been either removed from or redacted on the files from 1970.

In his submission, McMahon misleadingly advises Cabinet that Portugal had merely 'been making soundings with regard to a continental shelf boundary agreement with Australia in relation to the area between Australia and Portuguese Timor'. He does not explain Portugal's claim to the median line. He does not mention the Oceanic application.

On McMahon's recommendation, the Cabinet submission was approved by Prime Minister Gorton in mid-December without going to Cabinet.

As the February 1971 talks were confined to those areas where there was no dispute, agreement was quickly reached on all but a small section of the Indonesia–Papua New Guinea boundary. A median line in the Arafura Sea divided the area evenly and ended at the most westerly point that agreement could be reached in the Timor Sea.

McMahon replaced Gorton as prime minister in March 1971. The new foreign affairs minister, Leslie Bury, was briefed on Portugal's Oceanic permit in early April.[10] The brief does not mention Portugal's 'dispute' note.

Australia and Indonesia signed the first Australia–Indonesia Timor Sea Treaty on 18 May 1971. On the same day, Keith Brennan from the DFA met with Ambassador

Wemens and told him that there was 'no scope for negotiating a boundary since nature had already done that for us'.

NEGOTIATIONS, EXCHANGES AND
THE ARCHIPELAGO PRINCIPLE

Australian media coverage of the May 1971 treaty was not good for McMahon. Reports focused on the continued uncertainty about permits issued by Australia in the Timor Sea, and on the fact that if the median line principle applied in the Timor Sea a number of leases belonging to powerful international oil companies would end up in Indonesian or Portuguese waters.

Robin Ashwin, who had voiced concerns about the 'maximalist' position in 1965, was now at the Australian Mission to the UN. He took the unusual step of writing to his colleague Keith Brennan setting out his concerns.

> I doubt whether the value of any minerals we might extract from the area between a notional median line and the edge of the Trough in the next twenty or thirty years could be worth the risk we run ... of creating a substantial issue between us and Indonesia for the future.

Brennan thanked Ashwin for his letter. He attributed the 'Australian view ... that the Timor Trough separates the Australian continental shelf from that around Timor' to a 'recent and quite uncompromising reaffirmation by Ministers of the Government stand on the matter'. He then noted that: 'The fact that the Department of National Development believes the Timor Sea to hold particular promise for seabed exploitation makes any concessions in the area more than usually difficult.'[11]

Unknown to Ashwin in New York, his colleagues in Canberra were under escalating pressure. A surprise visit to Canberra on 6 July 1971 by a representative from Oceanic added to their angst. Wesley Farmer, vice president of geophysics, told Australian officials that Portuguese authorities had encouraged Oceanic to extend its exploration application to the median line.[12] The Woodside consortium had also sought approval to drill in the permit area that includes the Greater Sunrise field, and asked if 'it would embarrass the Australian Government in any negotiations it has in hand with Indonesia or Portugal if the Company were to drill at the point mentioned above'. Woodside asked 'whether there was any doubt of Australia's jurisdiction over the well site, and, if so, what was the source of the doubt'.[13]

It is obvious from the DFA files that there was good reason to doubt Australia's jurisdiction north of the median line in the Timor Sea. Portugal had formally contested Australia's right to issue permits. Geologists were supporting Indonesia's rebuttal of Australia's continental shelf claim and Indonesia continued to pursue a median line solution in the Timor Sea. And at the preliminary meetings for the Law of the Sea Conference there was growing support for a simple distance solution – a 200 nautical mile EEZ – to determine the limits of national jurisdiction over the seabed, which, as Hasjim Djalal, vice-chair of Indonesia's Law of the Sea Committee, pointed out to Australian officials, 'would produce a median line in the Timor Sea'.

Yet this uncertainty did not stop Australia advising Woodside on 25 August 1971 that its permits were valid and there were no issues with commencing drilling in the Timor Sea.[14] Remarkably, nor did this uncertainty stop Indonesia from agreeing to recommence seabed negotiations with Australia in early 1972.

An intriguing explanation for Indonesia's about-face emerges from the files in the archives. In the seventeen files titled *Australia–Indonesia Continental Shelf Boundaries Negotiations*, the documents concerning the May 1971 seabed treaty have nearly all been declassified. In contrast, the nine files in the series concerning the negotiations leading to a second treaty in October 1972 are heavily redacted. In 2016 the reason for withholding most of the material was because it contains information 'relevant to Australia–East Timor maritime boundary delimitation matters' that, if disclosed, could 'reasonably be expected to cause damage to the security, defence or international relations of the Commonwealth'.

However, there is also a small number of records

concerning the period from March to 9 June 1972 that remain classified for a different reason. These records are exempt because they 'contain information relevant to the capability, sources, objectives, methods, areas of interest or operations of an Australian intelligence agency', and that 'information is still regarded as sensitive'. On the advice of the DFA, the 'public disclosure of this information could compromise the future activities of an Australian intelligence agency and impair its ability to carry out its statutory functions'.

The use of this exemption indicates that some of the redacted material came from either the Australia Secret Intelligence Service (ASIS), an organisation so secret at the time that its existence was not publicly acknowledged; the Australian Defence Signals Division (now known as the Australian Signals Directorate), formed in 1947 to 'exploit foreign communications and be responsible for communications security in the armed services and government departments'; or the Joint Intelligence Organisation (now the Defence Intelligence Organisation), the intelligence arm of the Department of Defence.

In 1990 Brian Toohey and William Pinwill controversially published *Oyster*, a book based on a leaked copy of Justice Robert Hope's 1974 royal commission into ASIS. In a chapter discussing ASIS operations in Indonesia they recount Justice Hope describing a document obtained by ASIS on Indonesia's negotiating position for the Law of the Sea Conference as one of two ASIS 'diamonds'. By 'diamonds' Justice Hope meant 'pieces of information not otherwise or elsewhere obtained or obtainable which have proved to be of considerable significance to Australia'.[15]

Every time I came across a redaction in the files on the 1972 seabed negotiations (and there are many) I wondered if it could be hiding a reference to Justice Hope's 'diamond' – the intelligence that 'proved to be of considerable significance to Australia'.

At first it seemed odd that access to Indonesia's negotiating position for the Law of the Sea Conference would have been of particular value to Australia. Indonesia's key objective was no secret. At the 1958 and 1960 Law of the Sea conferences Indonesia had campaigned for international acceptance of the 'archipelagic principle', a radical concept that involved drawing straight lines linking the outer points of the islands in the Indonesian archipelago and declaring all waters within these lines to be internal waters. The concept was devised in the late 1950s by lawyer Mochtar Kusumaatmadja, known by all as Mochtar. He later recalled,

> The people had to be shown in simple symbols that Indonesia was one. We had just gotten our independence and we had all these big boys interfering, trying to keep us apart because they had their own designs. So this archipelago principle seemed to be a good thing for the important political unity of Indonesia.[16]

For Indonesia, maritime borders were both a deeply symbolic and a practical way to encourage a sense of nationhood in the 18,000 islands of the archipelago. But the archipelagic principle was anathema to Western naval powers, as it directly threatened the right of free passage between Indonesian islands, and it wasn't supported by the International Law Commission. If straits were classified as internal waters, permission from the Indonesian government would be required for transit. Australia, France, Japan, the Netherlands, New Zealand, the United Kingdom and the United States had issued diplomatic protests when Indonesia first asserted sovereignty over its archipelagic waters in 1966.[17]

The concept was particularly problematic for Australia, as it meant Indonesia claimed exclusive jurisdiction over some of the most travelled maritime routes between Australia and rest of the world. At the start of both the seabed boundary talks with Indonesia and the UN Law of the Sea preparatory talks, Australia opposed Indonesia's archipelagic principle, in step with its allies the United Kingdom and the United States.

While Indonesia's objective was hardly a secret, intelligence about just how important achieving international acceptance of the archipelagic principle was to Indonesia, and what Indonesia would be prepared to do to achieve that outcome, could have been 'of considerable significance to Australia'.

Indonesia's negotiations with Australia and at the

Law of the Sea Convention were led by Mochtar. As the 'father' of the archipelagic principle, Mochtar's main agenda was getting the conference to accept it.

It is clear from the declassified records in the archives that Australia softened its opposition to the archipelagic principle and then actively supported Indonesia's campaign at the Law of the Sea Conference. Exactly how 'intelligence' on Indonesia's negotiating position benefited Australia is impossible to judge, as many of the key files are still classified. However, a conversation Timor-Leste's foreign minister, José Ramos-Horta, had in 2004 with Indonesia's foreign minister, Hassan Wirajuda, is revelatory. Wirajuda told Ramos-Horta that Indonesia accepted Australia's continental shelf claims in 1972 not only because it was politically weak, but also in return for Australia's support for the concept of archipelagic states at the Law of the Sea Conference.[18]

THE AUSTRALIA–INDONESIA 1972 NEGOTIATIONS

During a visit to Canberra in January 1972, Mochtar delighted Australian officials with the sensational news that Indonesia would agree to reopen boundary discussions and offer a compromise boundary in the Timor Sea.

Soon after, Australian foreign minister Nigel Bowen offered to commence talks with Indonesia on the archipelagic issue and flagged that Australia wanted to take a middle path, in contrast to nations such as Japan, Russia and the United States, which wanted complete freedom of transit.[19] A new round of maritime boundary talks was scheduled for the middle of the year in Jakarta.

It can be gleaned from the declassified material that on 23 May, Bowen put a compromise line secretly negotiated by Mochtar and the DFA to Cabinet. It seems that Mochtar had agreed to a line halfway between the McCay line (along the bottom of the Timor Trough) and the median line – resulting in a boundary that essentially followed the northern limits of the permits issued by Australia.

However, while Mochtar and the DFA were in hearty (if secret) agreement, the Australian departments were not. The Department of the Interior, which was responsible for issuing permits, argued that Australia should insist on the Timor Trough/McCay line, because the 'permit areas allocated to Burmah/Woodside/Mid-Eastern, Shell, Arco/Aquitaine/Esso would be affected' by a compromise. It could not be 'assumed that an arrangement for Indonesia to take over parts of permits affected would be satisfactory to the Companies'.[20]

Bowen stressed that there would only be minor slivers, off a few permits, in areas that were of no known value, as a face-saving exercise for Indonesia. But Cabinet was not persuaded, and the 'maximalist' view prevailed. Cabinet directed the negotiating team to hold out for a boundary aligning with the Timor Trough.[21] The DFA was forced to cancel the Jakarta talks at the last minute, claiming more time was needed for 'important domestic considerations'.

On 29 August, in a Cabinet submission that is heavily redacted, Bowen again sought Cabinet approval for the DFA–Mochtar compromise. This time Cabinet endorsed the submission.

Talks were scheduled for early October. In preparation, Solicitor-General Robert Ellicott, who was leading Australia's delegation, asked for a map showing the location of wells drilled in the Timor Sea area. John Livermore from the Department of National Development circulated one and advised that

> You will notice that the wells known as Sahul Shoals, Eider and Flamingo are all well beyond the median line ... In addition we have marked the location of the Troubadour well which Burmah have had in contemplation for over a year but have not yet drilled.[22]

No wonder Australia embraced the compromise line: it skirted just to the north of the massive Greater Sunrise field – the hidden treasure the Australian government was determined to secure.

THE AUSTRALIA–INDONESIA TIMOR SEA TREATY

The negotiating teams at the October 1972 Timor Sea talks between Australia and Indonesia in Jakarta engaged in a highly choreographed exchange of compromise proposals until agreement was reached on a line halfway between the Timor Trough and the median line. At this point the Australian delegation cabled Canberra recommending acceptance because, 'as the Indonesians have reminded us on a number of occasions in the last two days, there is increasing international support for a simple 200 miles limit to national jurisdiction over resources'. The Australian delegation advised that the agreement would end the 'prolonged uncertainty which has inhibited companies from undertaking activity in some permit areas' and noted that it would be helpful to argue that 'the line, as now accepted, was essentially one proposed by Indonesia'.

More significantly, the cable concluded that the agreed boundary would 'place Australia in the best possible position when negotiating with Portugal in respect of the area south of Portuguese Timor'. As Paul Cleary notes in *Shakedown*, the location of the boundary points at either end of the gap left between Australia and Portuguese Timor caused

> the equidistance line to swing inwards towards East Timor and, incredibly, this happens exactly where significant resources are located. In the east the line swings in where the giant Sunrise and Troubadour fields are located, and in the west over the Buffalo, Laminaria and Corallina fields.

The deal was so good for Australia that rather than go through the usual process of getting Cabinet endorsement of the agreement, which could take some months, Foreign Minister Bowen immediately flew to Jakarta to sign the treaty. Bowen was advised to 'concentrate on the equal division of the disputed area' in his public comments, rather than the fact that Australia had secured 70 per cent of the disputed seabed between Australia and Indonesia in its jurisdiction. Indonesia appears to have paid a high price for Australia's support for the archipelagic concept.

There were, of course, other factors at play.

As parliamentary researcher Robert King details in a submission to a 2013 parliamentary inquiry, Australia engaged in a multi-pronged diplomatic pursuit of Indonesia in 1971 and 1972.[23] Australia boosted aid and defence spending in Indonesia and agreed on a 'favoured nation status' trade agreement. McMahon ignored opposition from the Defence Department and announced a gift to Indonesia of sixteen decommissioned Sabre aircraft as part of a new program of defence cooperation in parliament in December.[24]

Mochtar and Hasjim Djalal, vice-chair of Indonesia's Law of the Sea Committee (Mochtar was chair), have both cited a variety of reasons for Indonesia's capitulation in 1972: that Indonesia was unaware of the oil and gas potential of the region at the time; that Australia eventually convinced Indonesia that it was right about its continental shelf claim; that Indonesia knew it was right but couldn't produce sufficient evidence to prove its theory; and that Suharto wanted to be a good neighbour to Australia.[25]

Maybe it was a combination of all of the above.

One thing is certain: McMahon pulled off a diplomatic coup. Despite a baseless geological claim, and international moves towards a median line boundary in the Timor Sea, Australia secured most of the area it had unilaterally claimed in the early 1960s. McMahon also set up the scenario under which, should Portuguese Timor become part of Indonesia, it would be a simple exercise to close the Timor gap with a ruler, joining the end points of the 1972 treaty.

CHAPTER 3
AUSTRALIA VS. PORTUGAL: PRELUDE TO INVASION

Before McMahon could celebrate the diplomatic triumph of excluding Portugal from the Timor Sea talks or the economic triumph of the Australia–Indonesia Timor Sea Treaty, he was defeated by Gough Whitlam at the December 1972 federal election.

Whitlam set about reorienting Australia's relationship with Asia, particularly China and Indonesia. Before he became prime minister, Whitlam had met Suharto eight times.[1] Whitlam and his energy minister, Rex Connor, also had ambitious plans to use Australian energy resources to expand the industrial and manufacturing sectors of the Australian economy.

The Australian officials who had worked so successfully to secure the Timor Sea Treaty with Indonesia and keep Portugal at bay were now keen to complete boundary negotiations with Portugal as soon as possible.

Early in the new year the Bureau of Mineral Resources

provided the DFA with an 'assessment of the economic potential of the continental shelf opposite Portuguese Timor'.[2] In April Connor instructed the Woodside consortium to defer drilling 'pending ratification of the seabed boundary agreements with Indonesia, and negotiation of an agreement with Portugal'.[3] Senate government whip Justin O'Byrne referred to the pending talks with Portugal and enthusiastically told parliament that the Timor Sea was potentially one of the 'richest hydrocarbon empires in the world', containing 'gas and oil in quantities that could match even the fabulous riches of the Middle East'.

Australia's interest in the potential oil wealth of the Timor Sea escalated when the price of oil quadrupled following the cutback in production and embargo on oil exports to the United States and the Netherlands in October 1973 by the Organization of the Petroleum Exporting Countries (OPEC).

It was now Portugal's turn to stonewall. In November 1973 Portugal formally declined Timor Sea negotiations with Australia until after the next session of the Law of the Sea Conference, scheduled for mid-1974.[4] Like Australia and Indonesia, Portugal was keenly doing the numbers and would have been aware that the 200-nautical-mile EEZ proposal was likely to get up, which would vindicate Portugal's median line stance.

In light of Portugal's refusal to negotiate and the OPEC oil crisis, the DFA reviewed Minister Connor's decision to ban drilling in disputed areas of the Timor Sea and determined that 'Australia faces a serious situation regarding its own oil supplies, and Australian public opinion might well be critical of failure by the Australian Government to encourage efforts to locate and develop additional oil resources in Australia's offshore areas.'[5]

The DFA acknowledged the Portuguese believed the median line principle should be applied. However, it determined that a 'deterioration of relations with Portugal would not have serious consequences for Australia' and nor was 'Portugal in a position to impose its claims in the area by force. Moreover, Portugal has put itself in a morally weaker position by declining for ten months to discuss negotiations for settlement of the boundary.'

The DFA's advice concerning Indonesia noted that at some future date:

> Portugal might be obliged to withdraw from control of Portuguese Timor and control might then pass to Indonesia. As a potential inheritor, Indonesia thus has an interest in this seabed boundary. Indonesia, however, has given no indication that the drawing of a boundary line connecting the two extremities of the agreed Australian–Indonesian boundaries would be unacceptable to Indonesia. The indications if any, are to the contrary.

In other words, if Indonesia incorporates/invades/ takes over Portuguese Timor, we expect Indonesia will agree to close the Timor gap with a straight line, putting high-potential oil and gas areas on Australia's side.

DIPLOMATIC WAR BETWEEN AUSTRALIA AND PORTUGAL

Woodside was notified on 11 December 1973 that Connor had reversed his earlier advice and now agreed exploration could proceed in the Timor Sea.[6]

Portugal responded in kind, and on 31 January 1974 published a decree granting Oceanic exclusive rights to explore for and extract any oil, natural gas or other hydrocarbon products in a 14.8-million-acre area off the southern coast of Portuguese Timor.[7] Woodside translated all forty-seven pages of dense legalese and provided a copy to the Department of Minerals and Energy.

A DFA analysis of Oceanic's permit said the boundaries of the permit area

> followed closely the configuration of that median line ... in other words, the concession extends far south of the Timor Trough and the extrapolation of the agreed Indonesia–Australia boundary, and overlaps with parts of concessions granted to various companies by Australia and Australian states.

The Oceanic permit dramatically confirmed the doubts long harboured by Woodside and its partners Shell, Burmah Oil and BP, as well as the other Timor Sea permit holders, that Australia's jurisdiction north of the median in the Timor Sea was subject to challenge.

On 23 March 1974 Whitlam, who was acting foreign minister, approved a brief recommending Australia make an official protest to Portugal. In a note on the brief Whitlam queried if Woodside/Burmah should be informed.[8] Two days later, before the brief had made its way back to the department, Whitlam jumped the gun in a pre-recorded national television interview in Perth, announcing that 'just this last couple of days we have protested to Portugal for giving leases to an American company in our portion of the North-West Shelf which we had granted to Woodside Burmah ... These are very valuable resources.'[9] Following the interview, a member of the DFA public information office telephoned Canberra from Perth and passed on the gist of the prime minister's remarks. The interview was scheduled to go to air that night, causing panic in Canberra as the protest had not yet been issued.

It fell upon Acting First Assistant Secretary Hugh Gilchrist to telephone Portugal's Ambassador Wemens and ask him to call at the department as soon as possible. The ambassador arrived ten minutes later and Gilchrist, 'without even an Aide Memoire, had to ask the Ambassador to convey the protest to his government'.

Editorials in Australia's major newspapers suggest the DFA's public information office was busy spinning Australia's version of the dispute. The *Australian* claimed that on 'the basis of the Australia–Indonesia accords, our Government granted oil exploration rights to Woodside-Burmah. Portugal, which was not a party to those accords because it chose not to be, has granted rights in the same area to an American company, Oceanic.'[10] The *Australian* was wrong on two counts – the permits had been issued by Australia before the 'accord' with Indonesia, and Portugal's multiple requests to join the negotiations with Indonesia or negotiate bilaterally had been rejected by Australia.

The *Age* editorialised on the 'amicable and profitable agreement with the Suharto regime' and criticised the 'obstinate Portuguese' for refusing to 'come to the party'.[11] These articles contributed to a DFA myth that persists today: that the Timor gap was the result of Portugal's refusal to negotiate with Australia and Indonesia in 1972, a version of history that deliberately excludes the years between 1970 and October 1972 when Portugal tried on numerous occasions to initiate Timor Sea negotiations with Australia.

The archives reveal the DFA had no qualms about misleading parliament. During Australia's seabed negotiations with Indonesia in 1972, the ALP Opposition asked if Portugal was 'supporting Indonesia in seeking a median line between the Australian coast and Timor?' The question languished on the parliamentary notice paper until 26 October 1972, a week after the treaty was signed, when Australia's minister for development, Reginald Swartz, filed an answer. Ignoring the reality of the Oceanic permit application to the median line, the Portuguese dispute note of November 1970 and multiple attempts by Portugal to negotiate a median line boundary, the minister told the Australian parliament that the 'Portuguese Government has not made known its position'.

NEW PORTUGUESE REGIME SHIFTS THE DYNAMIC

Portugal formally responded to Australia's protest on 18 April. Its official statement noted there was one continental shelf in the Timor Sea, which meant the median line was the boundary separating the seabeds of each country.[12]

Before Australia had time to react, the dispute underwent a seismic shift. On 25 April 1974, the authoritarian regime in Portugal was overthrown in a peaceful coup. The new leftist government immediately announced Portugal would transition all Portuguese colonies to independence. No longer was Australia in conflict with a pariah imperialist state defying the United Nations' calls for decolonisation. Australia was suddenly in conflict with a progressive European nation representing the interests of the impoverished people of a potentially independent near neighbour.

We are now entering the timeframe covered by the Downer Compilation – the 484 documents from April

1974 to July 1976 concerning Australia's response to the Indonesian incorporation of East Timor that were selected for early release in 2000. It is apparent in the files I tracked through the archives that the dispute with Portugal was front and centre of 'Australian official thinking and action', to quote foreign minister Alexander Downer's explanation for the early release of the documents. Australia had spent over a decade defending the integrity of permits unilaterally issued in the Timor Sea in the early 1960s. Two years of intense diplomatic effort had gone into negotiations with Indonesia. There was a global oil crisis. Potentially billions of dollars were at stake in the dispute with Portugal over the right to issue permits north of the median line. Official protests had been exchanged.

Yet the scene-setting introductory essay in the Downer Compilation does not mention the dispute with Portugal, or the pressure the Australian government was under from the oil companies to close the Timor gap. A keyword search for 'Woodside', 'Oceanic', 'Burmah Oil' and 'BHP' draws a blank. A search for 'Shell' returns references to mortar shells falling in Dili.

One explanation for this silence emerges from the DFA's May 1974 report on the implications for Australia of the change of government in Portugal. The DFA concluded Portuguese Timor was economically unviable and, consistent with Cabinet's assessment in 1963, that it would inevitably be incorporated into Indonesia.[13] The report advised that Australia 'should beware of possible criticism of an attitude favouring the union of Timor with Indonesia lest it appear that prospects for the settlement of our sea-bed territory (and oil rights) dispute with the Portuguese were colouring our attitude'.

It is almost as if this advice was also applied to the editorial process for the Downer Compilation in 2000. As discussed earlier, the version of the May 1974 report reproduced in the Downer Compilation had an entire page of information about the dispute with Portugal, the Oceanic permit, and the concerns of the permit holders, omitted.[14]

However, in the archives, I found copious files on the dispute with Portugal, on Oceanic, on requests for certainty about the boundaries in the Timor Sea from the companies with Australian-issued permits, and the implications of the dispute for Australia at the Law of the Sea Conference. Needless to say, the discussion of Australia's Timor Sea oil agenda in the remainder of this chapter largely consists of documents that were not selected for publication in the Downer Compilation. Nor do the documents I discuss here tell the full story; hundreds of documents are still redacted and many files I have requested access to are still being assessed by the National Archives.

WOODSIDE STRIKES OIL

In June 1974, just over a decade after Boutakoff drew a box around an area of the Timor Sea much closer to Timor than Australia based on his reading of the contours of the seabed, Woodside finally manoeuvred the 'Big John' drill into place and tested Boutakoff's hunch. Woodside had been wanting to drill a well at Troubadour, in an area now part of the Greater Sunrise field, since at least September 1971. After numerous assurances from the Australian government that it should go ahead, drilling commenced on 4 June and good news soon followed.

The Troubadour well is much closer to Portuguese Timor than Australia. It is so close that Viqueque, a town on the south coast of Portuguese Timor, was used as the base for the drilling operation.[15] In Dili, where the local political elite were contemplating the nation's future, the throb of helicopters regularly making their way across the mountains to service the drill in the Timor Sea would not have gone unnoticed. In the circumstances, it was not unreasonable to consider an independent East Timor, backed by oil wealth from the Timor Sea, a viable proposition. The expectation, as Fretilin spokesperson José Ramos-Horta told the *Sydney Morning Herald*, was that 'eventually, Timor will be rich from oil but that will take time, just as independence will take time'.[16]

Expectations rose, particularly after oil and gas were discovered at the Troubadour well. Woodside and the Australian government, however, deliberately played down the implications of the discovery.

Professor Vivian Forbes, who worked closely with Woodside at the time, told journalist Hamish McDonald in 2013 that they 'didn't make a big thing about it. It was just a decision to make no big noise about the discovery as it could send shivers – shock waves – through the whole system.'[17] Canberra passed on news of the discovery to the DFA negotiators at the Law of the Sea Conference at Caracas with strict instructions to keep the news 'within the delegation'. In late August the *Age* reported Woodside had struck its 'first big gas flow off Timor', but Woodside insisted that other wells would have to be drilled on the Troubadour structure 'before the significance of the discovery was known'.[18]

Why all this secrecy in 1974? Two reasons. Confirmation of oil in the Timor Sea would have undermined claims an independent Portuguese Timor was economically unviable. And news of the find would have exacerbated Australia's Timor Sea dispute with Portugal. By then, Australian officials were aware that the Troubadour find extended west – which increased the likelihood of a discovery in the Australian permit areas directly off the coast of Portuguese Timor, overlapping Oceanic's permit.

This discovery, while difficult to access, was potentially worth billions of dollars to Australia. It is not mentioned in the Downer Compilation.

Woodside's concerns about a legal challenge from Oceanic or Portugal were now shared by the Department of Minerals and Energy. A declassified letter from the secretary of minerals and energy to the secretary of the attorney-general's department in August 1974 asks if, given the Oceanic 'overfile', drilling in the disputed area would still 'strengthen Australia's legal case in the event that the jurisdiction question' was considered by the International Court of Justice.[19] The answer was: 'Yes. Drilling ... would constitute the most concrete exercise to date of Australia's sovereign rights to explore and exploit the area.' Minerals and Energy then asked if Woodside's fears about possible legal action had substance and, if so, what would be the forum, and chance of success.

This time the attorney-general's department was more circumspect. Oceanic would only have hope of success in Portuguese courts. It was uncertain if Portuguese courts would have jurisdiction over an Australian company. In the circumstances, the department advised, it was reasonable for Minerals and Energy to insist the Woodside consortium meet its permit drilling commitments.[20]

SOWING THE SEEDS FOR INVASION

As Woodside was drilling for oil and gas, the seeds of Indonesia's invasion of Portuguese Timor were being sown.

One of the most explosive revelations when the Downer Compilation was released in 2000 was a cable Australia's ambassador to Indonesia, Robert Furlonger, had sent to Canberra on 3 July 1974. It linked Indonesia's plans for a clandestine operation to incorporate Portuguese Timor to a meeting Whitlam's private secretary, Peter Wilenski, had with Harry Tjan Silalahi from the

Indonesian Centre for Strategic and International Studies (CSIS). The cable reignited debate in 2000 about the extent to which Whitlam was responsible for shifting Indonesia's position from one of non-interest in Portuguese Timor to 'incorporation'.

The CSIS was a vehemently anti-communist think tank with links to Indonesia's secret service. In his autobiography, *The Hot Seat*, Richard Woolcott described the organisation as the most important source of accurate information on Indonesian policy and plans in relation to the future of Portuguese Timor.[21] The CSIS was run by Jusuf Wanandi, then known as Lim Bian Kie, and Harry Tjan Silalahi, the secretary-general of the Indonesian Catholic Party. Australian diplomats regularly met Wanandi and Tjan at the CSIS office in central Jakarta or over more intimate lunches and dinners at their homes.

In early July 1974 Ambassador Furlonger arranged for CSIS chief Tjan to meet Wilenski, who was holidaying in Yogyakarta. According to Wanandi's autobiography, the meeting was instructive:

> Peter explained his and Whitlam's thinking to Harry, namely that logically East Timor should become part of Indonesia. Doing so would ensure there was no lacuna that could otherwise be filled by another, potentially unfriendly power – a real fear in the Cold War climate, with North Vietnam close to victory over the South. It was after talking with Wilenski that

it occurred to Tjan that there might well be scope for more than a diplomatic initiative alone.[22]

Furlonger cabled Canberra, advising that 'Tjan's extreme frankness indicates that the Indonesians are confident that we would favour an independent Portuguese Timor as little as they do. (Tjan appears to have gained this impression from Wilenski.)' At Furlonger's suggestion, Tjan then visited Canberra in August to 'sound out' official thinking in Australia.[23]

We also know from the documents published in the Downer Compilation that Whitlam met with Suharto in Yogyakarta on 6 September 1974, accompanied by Woolcott and the secretary of the DFA, Alan Renouf. Whitlam advised Suharto that his personal view, which was likely 'to become the attitude of the Australian Government', was that 'Portuguese Timor should become part of Indonesia' and that this should 'happen in accordance with the properly expressed wishes of the people of Portuguese Timor'. Two weeks later, Whitlam amended his position: 'Obeisance has to be made to self-determination' so as not to 'create argument in Australia, which would make people more critical of Indonesia'. Whitlam's policy position did not contemplate a scenario in which the people of Portuguese Timor did not wish to become part of Indonesia.

The Downer Compilation also includes a cable from Lisbon quoting General Ali Murtopo's damning observation that until Whitlam's visit to Jakarta, Indonesia 'had been undecided about Timor'. The cable reports that Murtopo said Whitlam's 'support for the idea of incorporation into Indonesia had helped them to crystalize their own thinking and they were now firmly convinced of the wisdom of this course'.

The Department of Defence did not share the DFA's enthusiasm for incorporation. The *Age* reported on 13 September that Defence believed a friendly independent East Timor was in Australia's national interest.[24] Documents in the Downer Compilation confirm this view and also show that soon after Whitlam's Jakarta visit, Defence fell into line.

*

Whitlam's September meeting with Suharto led to another policy change – one not discussed in the Downer Compilation – that again concerned Australia's secret Timor Sea oil agenda. While Whitlam may have personally supported Indonesia's incorporation of Portuguese Timor, officially Australia continued to support the right of the Timorese to a vote of self-determination.

In the months leading up to Whitlam's meeting with Suharto, the DFA debated how to placate the Timor Sea permit holders and get Portugal to the negotiating table. Clearly unaware of the doubtful status under international law of Australia's unilaterally issued permits, in August 1974 Foreign Minister Don Willesee asked the

DFA to consider the merits of Australia taking action at the International Court of Justice against Portugal. The DFA's advice indicates the department was very aware that Australia's claim to permits north of the median in the Timor Sea was unlikely to stand up in court.

Gilchrist, who had delivered Whitlam's 'oral' protest to Portugal's ambassador, got the short straw again and had to tell Willesee that the idea was 'full of difficulties and uncertainties'. Gilchrist said, 'We would have to invoke some principle for delineation of the boundary but we departed from principle in order to reach a negotiated agreement with Indonesia regarding the adjacent seabed boundaries in the Timor Sea.' The DFA's formal advice back to the minister said:

> In any argument before the Court, Australia would muster considerable technical information to the effect that there is no common shelf. But the persuasiveness of such an argument, especially in the face of contrary views and relative uncertainty of the outer limits of continental shelves, may be regarded as an open question ... if the Court rejects Australia's argument that there is no shared shelf, then demands for renegotiation of the Indonesian agreements could not be discounted.[25]

Throughout August the departments continued to debate how to get Portugal to the negotiating table. By mid-August the Timor Sea boundary issue was being considered by the DFA 'in the light of Australia's relations with a future independent Timor' because 'moves for Timor's independence or self-government' were 'gaining ... steam'. Expectations of independence were so high that the DFA proposed including Timorese representatives in any seabed talks with Portugal.[26]

Australia's newly appointed ambassador to Portugal, Frank Cooper, was asked to assess likely Portuguese reactions to an approach from Australia 'seeking to open negotiations in the near future'. Cooper cabled Canberra on 30 August to advise that Portugal had no 'disposition to move on seabed boundary question at present' as 'this was a matter which may have to be left for decision by the Timorese themselves'.

This brings us to the eve of the Whitlam–Suharto September 1974 talks at which Whitlam expressed his personal support for an Indonesian takeover of Portuguese Timor.

Following the Whitlam–Suharto meeting, the DFA dropped the idea of approaching Portugal about Timor Sea negotiations. An interdepartmental meeting on 25 September 1974 agreed that, 'In the event of Timor's becoming associated with or part of Indonesia the Indonesians might well take the view that the Australia/Portuguese Timor boundary line should be no less advantageous to them than the existing Australia/Indonesia boundary line.'[27]

In other words, we won't waste time negotiating with Portugal (or the Timorese) as it is inevitable that Portuguese Timor will become part of Indonesia, and in that case, we expect Indonesia will agree to close the gap in the boundary on the same principles applied in 1972, i.e. with a straight line. This would of course, put all the known and potential oil and gas fields on Australia's side.

This scenario was flagged in an article on Whitlam's meeting with Suharto in the *Sun-News Pictorial*. Douglas Wilkie wrote, 'It suits both Jakarta and Canberra to tidy up a potential no man's land before it is exposed to big power subversion.' Wilkie then said Australia's acceptance of Portuguese Timor being merged into Indonesia will result in Indonesia looking 'favourably on Australia's search for off-shore oil in the area'.[28]

Soon after, Minerals and Energy and the DFA again advised that physical acts like drilling would be the best way to assert Australia's sovereignty in the Timor Sea.[29]

So for a brief period from October 1972 to August 1974 Australia was open to negotiating with Portugal to close the Timor gap. But following Whitlam's September 1974 meeting with Suharto, negotiations with Portugal were again off the agenda. This remained the case up to the Indonesian invasion of the territory in December 1975.

Around this time a representative of General Ali Murtupo consulted Woodside about its attitude to possible Indonesian occupation. The report in the archives does not include Woodside's response.[30] It does note that

Woodside was later advised to deal only with Indonesia's state-owned oil company Pertamina 'on matters concerning oil and gas exploration in the territory'.

Soon after the Whitlam–Suharto meeting, Indonesia commenced Operation Komodo, a covert campaign to achieve Portuguese Timor's incorporation into Indonesia. Operation Komodo focused on sowing confusion and tension inside Portuguese Timor. Indonesian agents across the border in West Timor beamed propaganda and disinformation suggesting that Fretilin was a communist organisation, and that support for integration with Indonesia was growing. These activities exacerbated tensions among the main parties in Portuguese Timor, and undermined the efforts of Portuguese authorities to develop a consensus around a plan for decolonisation. As Professor James Cotton notes, 'Without that campaign, differences between the indigenous political parties may never have developed so much as to lead to conflict.'[31] The Downer Compilation contains multiple references to diplomats from the Australian Embassy in Jakarta being briefed on the destabilisation campaign.

The Downer Compilation makes clear that not everyone in the government shared Whitlam's enthusiasm for Portuguese Timor's incorporation into Indonesia. It includes a letter Foreign Minister Willesee sent to Whitlam in January 1975 pointing out that self-determination was 'likely to yield a result other than the association of Portuguese Timor with Indonesia'.

On 21 February 1975 the *Sydney Morning Herald* published a story by journalist Peter Hastings that posited an Indonesian invasion of Portuguese Timor 'in the not too distant future'. There was a public backlash against Indonesia in Australia, a potent signal that that the broader Australian public did not share Whitlam's enthusiasm for an Indonesian takeover. Hastings' article sent the Australian bureaucracy into a panic. Whitlam and his DFA acolytes wanted Indonesia to take over Portuguese Timor – regardless, it seems, of how this came about. But the Australian public, the Opposition, and many members of the government, were outraged at the suggestion of a military invasion. Whitlam had to be seen to publicly protest the possibility, while at the same time assuring Suharto that he remained sympathetic to Indonesia's incorporation of Portuguese Timor.

By then the dominant public explanation for Indonesia's opposition to an independent Portuguese Timor was the need to stop the formation of an independent, unstable leftist state on its doorstep. The Downer Compilation includes a letter from Lance Barnard, the minister for defence, to Willesee in February 1975 in which he described Indonesia's concerns about the 'dangers of communist subversion' as 'unrealistic'. Barnard says, 'The principal factors stimulating the developments feared by Indonesia are the attitude and behaviour of Indonesia itself.'

In February 1975 Whitlam selected Richard Woolcott to replace Furlonger as Australia's ambassador to

Indonesia. Woolcott's foreign service career had already taken him to Russia, South Africa and Singapore, and he had served as Australian high commissioner in Ghana. In Canberra in the mid-1960s he had established the DFA's first public information office, giving him regular access to newspaper editors and influential journalists. Woolcott enthusiastically used the media to prosecute the department's interests throughout his career.[32] He was asked to draft Whitlam's first foreign policy remarks in government and accompanied Whitlam on key overseas visits. Woolcott was new school: a dapper, charming man with a drive to be modern, to be innovative, to encourage Australia to engage with its neighbours in South East Asia.

Woolcott personally delivered a letter from Whitlam to Suharto that noted the public debate in Australia about the Indonesia's alleged invasion plans and stated that no Australian government 'could allow it to be thought, whether beforehand or afterwards' that it supported resort to unilateral action.

OFFSHORE OIL EXPLORATION ESCALATES

The new government in Portugal had encouraged the formation of political parties in Portuguese Timor. Three main parties quickly formed around the obvious options for the future of the territory: the Revolutionary Front for an Independent East Timor (Fretilin) supported independence; the Popular Democratic Association of Timor (Apodeti) supported integration with Indonesia; and the Timorese Democratic Union (UDT), which by 1975 had formed an alliance with Fretilin.

Ambassador Cooper advised Canberra in early April 1975 that UDT and Fretilin would 'take a pretty tough line', in seabed negotiations while 'Apodeti would settle for a line matching 'that agreed with Indonesia.'[33]

In June 1975 the DFA sought advice about Portuguese-issued permits from Dr Avelar Barbosa, the head of Portuguese Timor's Inspectorate of Mines. According to Barbosa, Petrotimor (Oceanic in association with the government of Portuguese Timor) had 'all the off-shore rights beyond 200 metre depth, extending to the median line'. The DFA record of the meeting notes that Barbosa 'made it clear that he did not consider any other line to have any standing'.[34]

In Dili, talk of oil riches in the Timor Sea was renewed in July 1975 when Oceanic/Petrotimor established an office there and started a marine-seismic survey off the southern coast. Oceanic also caused a stir in Canberra in July. The company formally rejected Australia's claim to sovereignty north of the median line in the Timor Sea, advising the secretary of minerals and energy that the permits they held that extended to the median line were 'wholly consistent with the principles of international law in the Geneva Convention of 1957' and the principles enunciated at the recent Law of the Sea Conference which 'will probably be formally ratified at the next conference'. Oceanic bluntly stated that in the circumstances 'we fail to recognize the claim of your Government to this area'.[35] There is no reference to this letter in the Downer Compilation. There is, as we know, no mention at all of Oceanic.

The conflict over oil rights in the Timor Sea was escalating. And it was getting more complicated. The Australian company Timor Oil had exploration rights to the 200-metre depth line in the Timor Sea, under permits issued by Portuguese Timor. Woodside had taken a 70 per cent working interest in Timor Oil's concession in April 1974. By early 1975, the Timor Oil/Woodside Burmah group had completed an extensive offshore seismic survey. A well drilled 10 kilometres offshore in February 1975 caused 'frenzied trading' in Woodside and Timor Oil shares.

In July, Timor Oil was negotiating with Barbosa to extend its offshore exploration licence when Woodside suddenly pulled out of the venture, allegedly because of 'prohibitive costs' and a failure to attract partners. However, a DFA report later observed that 'the deteriorating security situation' and 'the possibility of Indonesian intervention, may have had a bearing on the decision'.[36] The timing of Woodside's withdrawal is remarkably prescient. As is well documented in the Downer Compilation, in August Indonesia's covert efforts to incite tensions between rival political groups in Portuguese Timor led to the pro-Indonesia integration parties instigating a coup in Dili that sparked a so-called civil war.

Between April 1975, when UDT President Lopes da Cruz and Vice President Costa Mouzinho visited Australia, and the outbreak of violence in August 1975, the UDT shifted its allegiance from Fretilin to the pro-Indonesia-integration Apodeti party. At a meeting with the DFA in Canberra in April, Da Cruz was still firmly in the independence camp, arguing that 'there was no need for a "referendum" on independence. It was already self-evident that a majority of the people of the territory already desired independence. One did not have to ask slaves if they wished to be free.'

During his visit to Australia, Da Cruz met with Bernard Callinan, a prominent right-wing Catholic and former commander of the Australian forces in Timor during World War II. Callinan was also a director of British Petroleum (Australia) – a major shareholder in Burmah Oil, one of Woodside's partners in the Timor Sea. In a pamphlet published in September 1975 the Sydney-based Campaign for Independent East Timor alleged that Callinan had advised UDT to break with Fretilin and form an anti-communist alliance with the pro-Indonesia Apodeti. Coincidentally or otherwise, by August the UDT had switched camp.

On 11 August 1975 UDT instigated a coup d'état in Dili, sparking a conflict with Fretilin. Up to 400 people are estimated to have died in the fighting. Barbosa was among 800 Timorese evacuated to Darwin. On his arrival, Barbosa told the Campaign for Independent East Timor that Timor Oil had 'continually stalled' the permit extension negotiations because, he believed, the company was 'waiting for a coup or invasion to renegotiate its leases as Indonesia gave much better conditions than the Portuguese, or Fretilin were likely to offer'.

Desperate pleas from the governor of Portuguese Timor, Lemos Pires, and Timorese support groups in Australia for assistance to stem the violence in Dili were met with ruthless silence from the Whitlam government. The government did respond, however, to Oceanic's letter rejecting Australia's claim to sovereignty north of the median line. In late August 1975 the secretary of minerals and energy wrote to Oceanic threatening 'substantial penalties' for breaching the terms of Australian legislation that 'prohibited exploration without a permit'.[37]

Apparently oblivious to the human cost of the violence in Portuguese Timor, on 17 August Ambassador Woolcott cabled DFA secretary Alan Renouf urging him to ensure the Australian prime minister did not do anything that could be construed as criticism of the Indonesian plan to incorporate Portuguese Timor. Woolcott then linked Australia's support for the Indonesian invasion with Australian oil interests in the Timor Sea.

> We are all aware of the Australian defence interest in the Portuguese Timor situation but I wonder whether the Department has ascertained the interest of the Minister or the Department of Minerals and Energy

in the Timor situation. It would seem to me that this Department might well have an interest in closing the present gap in the agreed sea border and this could be much more readily negotiated with Indonesia by closing the present gap than with Portugal or independent Portuguese Timor. I know I am recommending a pragmatic rather than a principled stand but this is what national interest and foreign policy is all about, as even those countries with ideological bases for their foreign policies, like China and the Soviet Union, have acknowledged.[38]

This is one of the rare occasions Australia's Timor Sea agenda was spelled out in black and white in declassified material. I suspect many similar statements on files in the archives are redacted on national interest grounds. I can't but wonder if Woolcott's advice quoted above would have been declassified in 2000 had it not already been published by Richard Walsh and George Munster in 1980, along with the full text of sixteen cables Woolcott sent to Canberra in July and August 1975. The publication of this paragraph in 1980 was the first indication from a DFA source that the exploitation of oil and gas in the Timor gap was an important consideration in Australia's response to the Indonesian invasion of Portuguese Timor.

There is little doubt that the prospectivity of the Timor Sea continued to interest the government. Following the Whitlam–Suharto meeting in September 1974, the Bureau of Mineral Resources had been asked for an estimate of expenditure on exploration in the area 'north of the median line between Australia and Portuguese Timor'.[39] In September 1975 the DFA asked the bureau for an examination of earlier assessments, noting that 'the area has significance with respect to future boundary negotiations'.[40] The sentence indicating the area in which the department was particularly interested is redacted, as is the entire second page attached to the request, as 'disclosure could compromise Australian sovereignty'. I suspect the fully redacted page is a map showing the area of particular interest. An annotation on the DFA file copy reads: 'Spoke to ... and stressed the importance the area may have on delimitation and asked for a thorough assessment.'

The revised assessment the bureau provided on 13 October 1975 has been fully declassified, as has an attached map that rates areas good, fair and poor, above and below the 200-metre depth mark.[41] It is odd that the request for information could compromise Australian sovereignty, but the information provided in response could not. The bureau's report clearly indicates the DFA was interested in the area 'opposite Portuguese Timor between the median line and the adjacent area boundary' – that is, the area overlapped by Portugal's permit to Oceanic. In the understated language of geologists, the paper concluded that that area had 'fair to good' prospects, particularly for gas.

PRELUDE TO THE INVASION

While officials in Canberra were assessing the oil and gas potential of the Timor Sea, five Australian-based journalists, Gary Cunningham, Brian Peters, Malcolm Rennie, Greg Shackleton and Tony Stewart from TV stations Nine and Seven in Melbourne, were travelling to Balibo, a small town in Portuguese Timor near the border with Indonesia. They were planning to report on rumours of a pending Indonesian invasion.

We know from the investigative journalism and scholarship of Jill Jolliffe, Hamish McDonald, Desmond Ball and others, and from the documents in the Downer Compilation, that the Australian Embassy in Jakarta was warned by their key contact Harry Tjan at the CSIS that Indonesia was commencing its Timor operation on 15 October. We also know that on 14 October Ambassador Woolcott had dinner in Jakarta with the architect of Indonesia's pre-invasion, incursion and subversion campaign, General Benny Murdani, who confirmed the plans outlined by Tjan.

On 16 October 1975 the five Australian journalists were shot or stabbed to death by Indonesian forces in Balibo. Their bodies were dressed in Fretilin uniforms, photographed and then burned. The Australian government officially denied knowledge of the journalists' fate despite intelligence that they had been killed. As Desmond Ball writes, a 'cover-up was instituted to ensure that the truth about Balibo never surfaced'.[42] In 2007 NSW Coroner Dorelle Pinch found there was 'strong circumstantial evidence' that Murdani gave the orders to kill the Balibo five.

The political situation in Australia meanwhile had reached its own crisis point. The Whitlam government was dismissed on 11 November, in part because of a scandal about the government's attempts to fund an ambitious network of pipelines to carry gas from the North West Shelf to the east coast.

Opposition Leader Malcom Fraser was sworn in as caretaker prime minister and a general election set for 13 December. In his autobiography, Fraser recalls that he was advised by the DFA that it was vital to Australia's long-term relationship with Indonesia that he send a message to Suharto indicating that a Fraser government would share the Whitlam government's policy on Indonesia.

> It really was put to me with the utmost urgency that it was most vital that I communicate to Suharto. On the one hand, I abhorred the incorporation. It was contrary to everything we believed about self-determination of peoples. On the other hand, I couldn't change policy anyway in caretaker mode and Indonesia was our most important neighbour, and I was being told it was vital to do this thing.[43]

It took a phone call with Woolcott to convince Fraser to authorise a top-secret cablegram to Suharto. The cable, strangely worded in the third person, indicated that Fraser 'recognizes the need for Indonesia to have an appropriate solution for the problem of Portuguese Timor'.

As Australian public opinion railed against the Indonesian invasion and demanded answers about the missing journalists, Fraser and his foreign minister Andrew Peacock were reminded by DFA officials that Suharto would see any criticism by the Australian government as a breach of faith and, in retribution, could leak Fraser's top-secret cablegram to Suharto, or the fact that the Whitlam government and the Opposition had been fully briefed on Indonesia's invasion plans.

In Portuguese Timor, fear of an Indonesian invasion intensified. Having secured control of the territory, Fretilin proclaimed East Timor's independence on Tuesday 28 November 1975. Xanana Gusmão was the official photographer at the solemn ceremony in the square between the Palácio Do Governo and the waterfront in Dili, where a crowd of 2000 people gathered at dusk to watch the raising of the Fretilin flag.[44]

Encouraged by Indonesia, two days later, the leaders of UDT and Apodeti signed the 'Balibo Declaration' over the border in West Timor, asserting East Timor's integration with Indonesia and calling on Indonesia to intervene. The same day, Gusmão left to volunteer to fight at the border, telling his wife not worry, that he would only be gone a few days.⁴⁵ It was nearly twenty-five years before Emilia and their two young children saw him again.

On 3 December Suharto met with his generals and gave approval for the Ministry of Defence and Security to take over East Timor.⁴⁶ In one of Woolcott's more jaw-dropping cables in the Downer Compilation, he advised Canberra to remind Peacock that Harry Tjan and Lim Bian Kie (Wanandi) believed that during a private visit to Bali in September, Peacock had said he

> favoured the early integration of Portuguese Timor into Indonesia ... He said that he would 'not criticise' Indonesia's actions to bring this about (including presumably the use of force) provided Indonesia had the support of other ASEAN countries ... you may need to drop the Minister a hint at some stage about what Tjan and Lim Bian Kie believe he said to them.

Australia's enthusiasm for an Indonesian-controlled East Timor was echoed by the world's superpower, the United States – but for different reasons. Indonesia was a prime market for American weapons and Washington saw Indonesia as a bastion of anti-communism in Asia. President Gerald Ford and Secretary of State Henry Kissinger met with President Suharto in Jakarta on 6 December 1975. Ford expressed 'understanding' about the 'problem' Indonesia had in East Timor, and Suharto agreed not to commence the invasion until they were back in the United States.

The invasion started at 2 a.m. on Sunday 7 December. Indonesian naval vessels shelled areas east and west of Dili, where Fretilin was believed to have stockpiled arms, while hundreds of paratroopers descended from the sky.[47] Gusmão describes the scene in his autobiography:

> We witnessed days of pillage through a huge pair of powerful binoculars. The weapons of war were vomiting fire over Dili's hills while cargo ships emptied the customs house of its contents ... An interminable line of people streamed upwards. I saw no fear in their exhaustion. I saw resignation in their eyes, and anguish that must have been torturing their souls, but they knew to smile, as if somehow it would relieve their suffering.[48]

Hundreds of Timorese were gunned down at random as Indonesian soldiers rampaged through the streets of Dili. At the wharf 150 people were shot, one by one, and their bodies dropped into the sea as the surrounding crowd was forced to count. Australian freelance journalist Roger East and Isobel Lobato, the wife of Fretilin leader Nicolau Lobato, were among those executed there. Around 400 people were killed in the first days of the invasion. Numerous other Fretilin supporters were rounded up, imprisoned and tortured.

A week following the invasion, Woolcott told journalists at a media briefing at the Australian Embassy in Jakarta that if Australia had helped in the formation of an independent East Timor, it could have become 'a constant source of reproach to Canberra ... It would probably have held out for a less generous seabed agreement than Indonesia had given off West Timor.'⁴⁹

CHAPTER 4
DEATH, DENIAL AND OIL

The UN continued to recognise Portugal as the official administrative authority in East Timor following the invasion. On 12 December 1975, the day before the Australian election, caretaker foreign minister Andrew Peacock instructed Australia's delegates at the UN to support a resolution calling on Indonesia to 'withdraw without delay its armed forces from the territory in order to enable the people of the territory freely to exercise their right to self-determination and independence'.

Despite a media lock-out imposed by Indonesia, it soon became clear that the Timorese were putting up major resistance. Reports of a rising death toll and atrocities made it into the Australian media based on interviews with refugees, letters smuggled out of the territory, or relayed via Fretilin's one radio link to the outside world in Darwin. Yet just three years after the invasion, the Fraser government became the only Western country to fully recognise Indonesia's sovereignty in East Timor.

The DFA-approved explanation for Australia's decision to recognise Indonesia's sovereignty is that because the Whitlam government had secretly encouraged Indonesia's occupation of East Timor, the Fraser government was obliged to support Indonesia in the aftermath of the invasion at the UN, and this extended naturally to recognising Indonesia's sovereignty. The Fraser government's hands were tied.

There is some truth to this. But it is a convenient truth. As I continued to track Australia's claim to oil-rich areas of the Timor Sea oil through the archives, another, uglier story emerged. A story that is difficult to write.

The expectation of the DFA (based on the advice of the Indonesian generals) was that the military engagement would be over quickly and the Timorese would come to welcome being part of Indonesia. As discussed earlier, Mochtar and others had also led the DFA to expect that following East Timor's successful integration, Indonesia would agree to close the Timor gap with a straight line, joining the end points of the Timor Sea Treaty. This would finally provide the oil companies with the security of tenure necessary to commence drilling and the commercialisation of fields like Woodside's Greater Sunrise.

But the mandarins in the DFA had not prepared for a scenario in which the Timorese fought back and resisted the Indonesian occupation.

THE FRASER GOVERNMENT'S FIRST ATTEMPT AT NEGOTIATIONS

By the end of December, Timor Oil and other companies with permits issued by Portuguese Timor had reached a 'suitable understanding' with Indonesia, and agreed to delay exploration without protest in return for a guarantee from the Indonesian government of their present positions in the future.[1]

Oceanic was the exception. Three days after the invasion, Oceanic sent a letter to the Australian government advising that they intended to carry out geophysical and eventual drilling activity in the Timor Sea as soon as the political situation in Timor 'normalised'.[2]

Mochtar, who was now Indonesia's minister for justice, made an early play to get the Timor Sea boundary issue under discussion.

The 1971 and 1972 treaties had only determined a seabed boundary. Australia and Indonesia still needed to negotiate a 'water column' or fisheries boundary. During a lunch with Woolcott in late March 1976, Mochtar offered to negotiate a fisheries boundary based on the seabed line. Woolcott's response conflated the need for a fisheries boundary between Indonesia and Australia with the closure of the Timor gap. He said he thought the seabed treaties were 'very satisfactory' and that he 'had always hoped' that Australia's 'boundary could be completed by filling the gap south of eastern Timor'.[3]

In Canberra the DFA found Mochtar's offer 'puzzling'. There was concern that the 'appearance of actual disadvantage to Indonesia in negotiating such a fisheries boundary could' lead to 'attack on the grounds that it grossly favoured Australia' and therefore 'place the seabed boundary at risk'. It was too good to be true.

Nevertheless, when Peacock met with Mochtar two weeks later in Jakarta, he asked if the idea of a fisheries boundary matching the seabed boundary was a 'firm proposal'. Mochtar confirmed that it was, saying, 'It was not worth worrying over a few miles.' Peacock then raised the need to negotiate a seabed and fisheries boundary with the provisional government of East Timor – the Indonesian administration supposedly in control in East Timor – noting that if Timor decided 'to integrate with Indonesia, the negotiations would presumably take place with the Indonesian Government'.[4]

This exchange was not included in Woolcott's notes of the meeting approved for release to the Australian media. The media notes did record, however, that Mochtar confirmed the existing seabed treaties would not be affected by the outcome of the Law of the Sea Conference and Peacock confirmed Australia's in-principle support for the archipelagic principle.[5]

All that was needed now to close the Timor gap was East Timor's successful integration into Indonesia.

On 17 July 1976, following a sham self-determination process, Suharto signed a bill declaring East Timor

Indonesia's twenty-seventh province. Reports of the brutality of the invasion had appeared in the Australian media and community backlash was growing. The issue was considered so controversial in Australia that Indonesia agreed to bring the date forward from August to July to ensure the Australian parliament would not be in session.[6]

Peacock ignored advice from Jakarta and publicly said that as the UN had not been involved, Australia did not regard 'the broad requirements for a satisfactory process of decolonisation as having been met'. The Australian media criticised Peacock for not going hard enough against Indonesia. The Indonesian media gave the statement wide coverage and saw it as a rebuff.

Peacock's statement totally derailed Australia's Timor Sea oil agenda. If Australia didn't recognise Indonesia's sovereignty in East Timor, then it could not negotiate a treaty with Indonesia concerning East Timor. Which meant it couldn't close the Timor gap.

Peacock's statement also derailed Indonesia's agenda, which was to get Australia to recognise Indonesia's sovereignty in East Timor. As an Indonesian general explained to Woolcott, 'Australia's position was especially important because it was a close neighbour and ... its views carried weight in the region.'[7]

A visit Fraser was planning to Indonesia in October presented an opportunity for Australia to get negotiations to close the Timor gap back on track. Peacock, who would accompany Fraser, was briefed on 'formulas for recognition' by the acting deputy secretary of the DFA, Alf Parsons. Peacock was offered a choice of 'overt' and 'covert' ways to recognise Indonesia's sovereignty. The overt option involved a simple public statement of recognition. The covert option involved the prime minister giving a private assurance to President Suharto that Australia recognised the integration but explaining that he 'could not say so publicly for the present'. Peacock wrote on the brief: 'No – under no circumstances should a covert formula even be contemplated.'

In a discussion of 'next steps' Parsons set out what proved to be Fraser's strategy (despite Peacock's 'No' next to the paragraph on the brief). Parsons wrote:

> In short, we could be planning and hoping for a gradual fading away of the problem ... until we reach a point where we can slip into a form of recognition without having to say so publicly (e.g. an official visit to East Timor by the Ambassador or one of his senior staff).[8]

Underlying Parsons' approach is the distinction in international law between *de facto* and *de jure* recognition of states. *De facto* recognition means a government accepts that another government has effective control of a territory even though that control is not formally or legally recognised. This is a pragmatic, lesser form of recognition, necessary to conduct basic relations with another state. *De jure* recognition means a government accepts the legal validity of another government's sovereignty. This is the highest form of recognition, necessary to enter treaties and engage in other high-level diplomatic activities like official ambassadorial and state visits.

Peacock's response to Indonesia's declaration in July meant Australia had not recognised Indonesia's occupation of East Timor on either a de facto or de jure basis.

Fraser planned to offer de facto recognition during his October visit to Indonesia, and may have been willing to go further. Peacock argued it was too soon after Australia's refusal to accept Indonesia's act of integration.[9] At the last minute, Fraser deleted a line about recognition from his speech to the Indonesian parliament.

It was a pyrrhic victory for Peacock as Fraser then employed the covert strategy Peacock strongly opposed, telling President Suharto during their private meeting that Australia would recognise Indonesian sovereignty de facto but couldn't say so publicly yet.[10]

These events have been extensively analysed by historians, international lawyers and political biographers. A fascinating, almost contemporary account by Whitlam's former foreign affairs adviser Nancy Viviani was published in 1978. In her paper, based on publicly available information at the time, she attributes Fraser's moves to seek an 'accommodation with Indonesia' in October 1976 to media reports that Australia had been warned by high-ranking US officials 'not to allow any further deterioration of relations with Indonesia'. America wanted a 'friendly anti-communist' government next to the deep-water Ombai and Wetar straits used by United States and Soviet nuclear submarines.[11]

This was a reference to a Fairfax story in August, just after Peacock's refusal to accept Indonesia's act of integration. The Ombai and Wetar straits may well have been of primary concern to the United States, but I doubt they were the primary factor motivating Australia's moves to recognise Indonesia. I wouldn't be surprised if the leaking of such an otherwise sensitive defence issue to the Australian media in South East Asia, where the journalist Michael Richardson was based, was facilitated by the Australian Embassy in Jakarta to add to the pressure on Peacock to support Indonesia's occupation.

I suspect the primary factor motivating Australia's moves to recognise Indonesia was determination to close the Timor gap. In the various accounts of Fraser's visit there was no mention of the maritime boundary issue. Yet while Fraser was meeting with Suharto in Jakarta, Australian officials opened negotiations for a seabed boundary between Australia and East Timor. The talks were covered on the front page of the *Australian* under the headline, 'Now for talks on seabed rights'. The story stated:

> The aim is to clear the way for Australian oil explorers in the area. The approach in effect acknowledges Indonesia's sovereignty over East Timor and underlies

Australia's apparent acquiescence to the annexation of the former Portuguese colony. The discussions were held on an official-to-official basis while the Prime Minister, Mr Fraser, was holding talks with President Suharto.[12]

The other story that greeted Fraser on his return to Australia, clearly emanating from Indonesian sources, claimed he had accorded de facto recognition to East Timor's integration into Indonesia during his visit to Jakarta. This led to Fraser facing a no-confidence motion in parliament. During the debate Whitlam referred to 'the pressure being brought to bear on the Australian Government to negotiate with Indonesia concerning the petroleum exploration permits in the portions of the seabed closer to East Timor than to Australia'.[13] Fraser ignored the seabed issue in his response and instead linked the need to engage with Indonesia about East Timor to refugees and family reunions – the humanitarian ruse – and attacked Whitlam's record on Timor.[14]

A week later the *Canberra Times* headlined a story 'PM accused of "illegal" talks on sea border'.[15] The article quoted a Fretilin spokesperson stating that as Australia

does not recognise the Indonesian takeover of East Timor, then it follows that such talks are illegal and contrary to the wishes of the East Timorese people ... Negotiations with Indonesia on the subject would

> amount to an acknowledgement by Australia of
> Indonesian sovereignty over East Timor – a
> diplomatic step, which up to now, it has not been
> prepared to take.[16]

The furore over the recognition issue and Timor gap negotiations led to the talks being abandoned. Mochtar responded to the cancellation by raising the stakes. He told Australian journalists that if Australia acknowledged Indonesian sovereignty by negotiating with Jakarta to demarcate the sea boundary in the Timor gap, then Indonesia was prepared to give Australia the same favourable deal it had conceded in the 1972 seabed negotiations.[17]

Australia's attorney-general was now Robert Ellicott, who had led Australia's negotiating team at the October 1972 seabed talks with Indonesia. He was forwarded a copy of Mochtar's comments with the advice that seabed talks 'would, of course, involve recognition of Indonesia's sovereignty over East Timor'.[18]

KILLING THE MESSENGER – THE DUNN FILES

The response of the Kissingerian realists in the DFA to the recognition–Timor gap negotiation nexus was to attempt to create a domestic environment in which it would be politically acceptable to recognise Indonesia's sovereignty. What was needed was Indonesia's successful integration of East Timor, no more stories of atrocities in

East Timor in the Australian media, and the issue to drop away at the UN.

Academic Clinton Fernandes, a former Australian Army intelligence officer, has examined records in the National Archives of Australia that reveal that 'Australian diplomats and other government officials were aware that a major humanitarian catastrophe was occurring in East Timor'.[19] Australian intelligence agencies regularly monitored Indonesian military communications and would have briefed senior officials and politicians on these events. The records that would show the full extent of the Australian government's knowledge of the famine are redacted, like many Timor Sea oil records, on the grounds their release would compromise Australia's security, defence or international relations.

As I continued to work through the Timor Sea oil files from 1976 and 1979, I discovered a disturbing link between the cover-up of Australia's knowledge of atrocities in East Timor and Australia's Timor Sea oil agenda.

The Australian Embassy in Jakarta had two fat files concerning reports by former DFA diplomat James Dunn in 1976 and 1977 about a rising death toll and atrocities in East Timor.[20] Dunn was a defence specialist focusing on Indonesia, and from 1962 to 1964 Australia's consul in Portuguese Timor. He'd served as the West European desk officer followed by postings to Paris and Moscow before becoming director of the Foreign Affairs Group of the parliament's Legislative Research Service in 1970. He was seconded back to the DFA in 1974 at Whitlam's request to report on the situation on the ground in Portuguese Timor. His report recommended self-determination, putting him out of favour with Whitlam.[21]

He was soon also out of favour with the Fraser government. A month after Fraser's bungled Jakarta visit, a report from Indonesian Catholic Church sources was smuggled out of East Timor and given to Dunn. The report estimated that between 60,000 and 100,000 Timorese had been killed since Indonesian forces had invaded. The allegations were published in Australian newspapers and Dunn sent a copy of an analysis he prepared for MPs to his former colleagues at the DFA.[22] In a covering note Dunn observed that 'Indonesia may have committed one of the worst atrocities, relatively speaking, in modern history in East Timor'. An annotation on the copy of Dunn's note on the file from the Australian embassy in Jakarta says: 'Everything is relative my dear chap!' Other annotations on the report are whited out.[23]

The Dunn files in the archives are not compilations of attempts to pressure Indonesia to investigate the allegations, or analyses of diplomatic options available to Australia to express concern about the seriousness of the allegations, or of government statements condemning Indonesia's treatment of the Timorese.

The files are compilations of the DFA's attempts to downplay, dispute and dismiss Dunn's allegations.

It wasn't just Dunn's allegations that got this

treatment. In November Canberra sent a media release issued by Fretilin to the embassy in Jakarta. The second paragraph states: 'In the Baucau concentration camp the enemy are daily torturing, raping and executing the captured population.'[24] Next to the underlined text, in what Cavan Hogue, Woolcott's deputy, told journalist Tom Allard in February 2016 'does looks like my handwriting', are the words 'sounds like fun' and the comment that: 'This report is internally inconsistent. If the enemy was impotent, as stated, how come they are daily raping the captured population? Or is the former a result of the later?' The recipient, another senior Australian diplomat in the embassy, has written 'sounds like the population must be in raptures' and notes that, 'I think that for our records we ought to ask the Hague to send us more [Fretilin reports]. In any case they'd make a great cure for constipation.'[25]

Peter Rodgers, the embassy's first secretary, defended the annotations, telling Allard that:

> Those in the embassy in 1976 had no more reason to believe Fretilin propaganda than they did to believe Indonesian, UDT, Apodeti propaganda over the situation in East Timor. The commentary was blunt but this was on claims made by one of the protagonists in a messy, propaganda-rich, conflict.

This comment is surprising given the embassy's Indonesian sources were, if anything, verifying the death toll claims. For example, when Allan Taylor, the political counsellor at the embassy, asked Indonesian General Sunarso about the alleged casualty figure of 100,000, the general gave the impression 'that he did not think 60,000 beyond the realm of possibility'.[26]

In response to the death toll allegations in the Catholic Church report forwarded by Dunn, Woolcott arranged another lunch with Murdani. In a long cable to Canberra, Woolcott reported Murdani's claims that 'much worse things' were happening in Cambodia, Laos and the Philippines, and that the figure of 100,000 deaths was 'ridiculous', and 60,000 'certainly exaggerated'.[27] Woolcott also passed on the general's view that Australia's request not to deploy the gifted Sabres to East Timor 'showed Australia's potential lack of resolution and unwillingness to support Indonesia'. While he acknowledged he had limited capacity to check the accuracy of Murdani's claims, Woolcott said that Murdani gave the impression that he was not worried about the situation in East Timor and that 'given his responsibilities at the policy level and his personality I think it would have shown if he had been concerned about it'.

What exactly was Murdani not concerned about? The bad media? The lack of Australian Sabres? The rising death toll? Woolcott does not elaborate.

Meanwhile in Canberra the DFA was still having to deal with increasingly unhappy Timor Sea permit holders. In January 1977, a representative from Shell (a major partner in Woodside's Greater Sunrise venture) sought a meeting with Peacock to discuss Australia's willingness to negotiate with Indonesia about 'closure of the gap'. By then Australia's biggest mining company, BHP, had bought Burmah Oil's share in the North West Shelf venture. As a result of an attempted takeover, Shell and BHP now had a combined 40 per cent share of Woodside. The DFA advised Peacock that

> because of the political sensitivity of the issue, a decision to negotiate and the timing of negotiations will have to be handled carefully, as such negotiations would of course constitute de jure recognition of Indonesian sovereignty over East Timor. If in due course negotiations are commenced, the best solution for Australia would be a relatively straight line closing the present gap.[28]

This advice confirmed that Australia would have to give not just de facto recognition to Indonesia's occupation of East Timor, which was proving difficult enough, but would have to acknowledge Indonesia's full de jure sovereignty.

While the DFA was crafting a way forward, James Dunn was in Portugal interviewing 200 of the 1500 Timorese refugees in Lisbon on behalf of Australian Catholic Relief and Community Aid Abroad. His report was published in the Australian media in February and proved to be another major setback for the DFA's recognition agenda.

Dunn described the brutality of the initial invasion, the killings on the Dili wharf and claims by a priest that 'the entire Chinese population of Maubara and Liquica was shot by Indonesian troops when they entered these villages'. He reported that 'refugees became prisoners in the refugee camps and none of the aid channelled via the Indonesian Red Cross reached the refugees'.

The DFA was quick to disown Dunn. Allan Taylor advised his British colleagues that the report was written in Dunn's 'private capacity and it had no official status. He was not a member of the Department'. The 'allegations would be difficult, if not impossible, to authenticate' and had been 'exaggerated'.[29]

I found an annotated copy of Dunn's report on a Jakarta embassy file. The nadir of annotations is next to a statement that there had been a 'great deal of looting and raping of girls in Baucau'. Someone has written: 'How do you loot a girl? (misspelling?)'[30]

> There were few first hand accounts of the situation outside the Balibo/Maliana and Dili areas. One of these was from a Timorese who said that he was in Baucau when the Indonesians attacked that town on 10 December 1975. He said there were few casualties at first, but that later, the troops shot many Chinese. There was, however, according to this informant a great deal of looting and raping of girls in Baucau. I was also told that fighting in the Baucau/Los Palos/Viqueque area had last year been very intense, because of effective operations by Fretilin.

The embassy in Jakarta was not alone in criticising the messenger. Bob Santamaria, a prominent right-wing Catholic, falsely claimed Dunn was a committed supporter of Fretilin, leading a 'campaign against Indonesia'. Conservative Australian National University academic Heinz Arndt publicly attacked Dunn and 'the left' for turning the media and the public 'against Indonesia'.[31]

Dunn's report led to calls for a federal parliamentary select committee inquiry into the deaths of the Balibo journalists. When the Fraser government failed to support the inquiry, nearly 100 federal parliamentarians from both political parties signed a petition in February 1977 to US president Jimmy Carter about alleged Indonesian atrocities in East Timor.

In a powerful opinion piece in the *Australian*, government MP Michael Hodgman said Dunn's recent allegations 'cry out for prompt and painstaking investigation, either by a United Nations mission ... or by a visit to East Timor by an Australian parliamentary delegation'. Unlike the DFA diplomats, Hodgman appreciated the horror of what Dunn reported:

> For us, as Australians, to bury our heads in the sand and turn our backs on what is alleged to have occurred, would be a gross act of national moral cowardice. We would be degrading Australia, and future generations would have to bear the same shame and disgrace which fell upon those citizens

> who turned a blind eye to Auschwitz by the simple
> process of saying to themselves: 'It does not exist – it
> has not occurred.' The dilemma of East Timor will
> not go away – it will not conveniently disappear. The
> ghosts of the dead will haunt each and every one of
> us who seeks solace in silent acquiescence.[32]

In March the *Age* ran a front-page headline: 'Indon Threat on Timor, Canberra Told "Keep Dunn Quiet".[33] Dunn did not go quiet. With the Australian government ignoring his report, he went to the United States and testified before a committee of Congress and gave evidence to the Fourth Committee of the UN General Assembly in March 1977. Meanwhile, individual members of Australia's parliament continued to challenge the government's version of events in East Timor, and were met with half-truths and outright lies.

Ken Fry, who had been a member of a parliamentary delegation that visited East Timor in September 1975, used the parliament's Joint Committee on Foreign Affairs and Defence to quiz DFA officials about Australia's Timor Sea boundary negotiations. The declassified transcript of a closed hearing on 30 March reveals the DFA's expertise at obfuscation.[34] The DFA officials presented a paper outlining the history of the Timor Sea boundary dispute that stated that: 'At various times since 1971 Portugal and Australia discussed possible negotiations to close this gap,' artfully failing to state that Australia had refused

Portugal's requests to negotiate. In response to a question about whether the Portuguese government or Oceanic consulted Australia before the permit to Oceanic was issued, they said, 'No, not in advance or subsequently.' This ignored Portugal's numerous attempts to consult with Australia about the risk of overlapping permits after being approached for a permit by Oceanic.

The officials' response concerning Indonesia, however, accurately reflected the ongoing expectation that Indonesia 'would be prepared to simply join the 2 ends of the line and close the gap by almost a straight line'. They also confirmed that the government was under pressure to provide certainty for the oil companies.

The DFA also had to deal with the loose lips of Indonesian ministers. The front page of the *Canberra Times* reported Foreign Minister Malik's comment on the death toll in East Timor that: 'The total may be 50,000, but what does this mean if compared with 600,000 people who want to join Indonesia? Then what is the big fuss?'[35] Woolcott, unhappy that Malik had 'stirred up the question of the number of people who had been killed in Timor', asked Murdani 'for his frank assessment of casualties'. This time Murdani said that 'in his view the total killed, including 745 Indonesians killed from August until now, would not exceed 7000 at the outside'. Woolcott also uncritically relayed to Canberra Murdani's repeated claim that the Western media unfairly attributed all the deaths to Indonesia.

EDGING TOWARDS RECOGNITION

In early April 1977 the DFA's General Legal Section stated the obvious: 'Indonesia's acquisition was illegal.'[36] The advice also argued that the Australian government had acknowledged Indonesia's de facto sovereignty 'by negotiating with the Indonesian Government on the humanitarian and refugee issue (irrespective of whether this had been admitted publicly)'. In which case, a visit by senior Australian diplomats to East Timor to check on aid deliveries, hosted by Indonesia, would help cement de facto recognition as Parsons had earlier advised.

Woolcott arranged for Hogue and Rodgers, the embassy officials discussed earlier in relation to the annotations, to visit East Timor for three days in April 1977. The file covering their visit makes fascinating reading. Every minute of their trip to report on the use of $333,000 of Australian aid through the Indonesian Red Cross and to prepare for a visit of an Australian immigration team to facilitate family reunions is stage-managed. They are accompanied at all times by their Indonesian hosts, followed by a large Indonesian media pack and met by 'protestors' supporting integration with Indonesia at various locations.

Their report notes that they assumed their hosts were 'trying to hide some things from us and that they were trying to make a case that we would believe'.[37] 'Everywhere we went we were shown crates of material which had signs on them saying "Contribution from Australia" it is doubtful if this bore a close relationship to reality.' They were transported around the island by helicopter and their flight path regularly diverted out to sea. Hogue later conceded that the explanation given by an Italian priest that this was in order to avoid areas controlled by Fretilin 'may well be right'.[38] He also noted that the priest's assessment of the behaviour of the Indonesian troops was consistent with his observation that 'the Indonesians tended to look on a stint in East Timor as an exile from civilisation'.

Photos in the archives left me in awe of the *Wag the Dog* cynicism of their visit. Among the pictures of unsmiling school children in pristine uniforms doing exercises, or desultorily waving Indonesian flags, I came across photos of banners on buildings, or strung between trees, in English, Indonesian and Tetun. 'James Dunn go to hell', 'Jim Dunn = trouble maker', and 'I love the Australians but more integration'.[39]

Who was the intended audience – the Australian diplomats, or the Australian public? If the latter, Dunn's notoriety was so obviously confected that the photos never saw the light of day in the Australian media.

A story on Hogue and Rodgers' visit did, however, make the front page of the *Age*. The opening line of the article stated that 'East Timor's integration with Indonesia appears "to be clearly an irreversible fact," according to a report to the Federal Government.'[40]

Keeping on message, in May Woolcott advised Nick Parkinson, who had replaced Renouf as secretary of the DFA, that the embassy's 'earlier assessment that integration was an established fact and was irreversible' was supported by the recent Hogue–Rodgers visit.[41] In the same cable, Woolcott complained that the advantages gained by Fraser's visit to Jakarta in October 1976 had been partially dissipated by 'government statements of non-recognition of incorporation, through the activities of some members of Parliament and by situations such as that stirred by Dunn'. He then directly referred to the link between the Timor Sea boundary negotiations and recognition, arguing that closure of the gap in the seabed boundary in the Timor Sea could not be negotiated with Portugal. 'It must be negotiated with Indonesia. This can presumably only be done on the basis of our recognition of Indonesian control of East Timor. The longer the question is left unresolved the harder it might be for us to achieve a suitable agreement.'

Parkinson was unmoved. He thanked Woolcott for his views and told him frankly that the government was not prepared to recognise Indonesian sovereignty.[42]

Soon after, Oceanic made a strategic move that panicked the Australian permit holders. The company announced it was seeking Indonesian endorsement of its permit granted by Portugal. The *Australian Financial Review* reported that 'close observers of the Indonesian scene believe Jakarta is deliberately reserving judgment on the matter as part of a campaign to try to accelerate a formal acknowledgement from the Fraser Government that East Timor is now Indonesia's 27th Province'.[43]

The Australian permit holders intensified their lobbying campaign. French company Aquitaine had the permit containing the highly prospective Kelp area in the centre of the Timor gap which was overlapped by Oceanic's permit. In late May, Aquitaine sought to defer its drilling obligations on the basis that a large discovery before resolution of the boundary question could 'exacerbate the present boundary problems and considerably delay their final settlement'.[44]

In early June, the minister for natural resources, Doug Anthony, sought advice from the DFA on the 'prompt settlement of the East Timor seabed boundary question'.[45] Ian Sinclair, acting foreign minister, responded on 23 June 1977 acknowledging the minister's 'particular concern to respond in some definite sense to the companies which hold the relevant offshore exploration permits'. Sinclair wrote that

> Indonesia has assumed responsibility in the territory and there is no reason to believe this situation will change. The fact remains however that the Australian Government has not yet felt itself able to take action, such as negotiating a seabed boundary south of east Timor, which Indonesia would be able to characterize as recognition of Indonesian sovereignty over east Timor.[46]

Woodside kept up the pressure, complaining in a letter on 29 June that the consortium's concerns about the 'security of tenure of the permit have not been alleviated'. Which meant that despite the consortium's 'keen interest in an attractive prospect in the permit' they were not prepared to commit the funds to drill a well.[47] The letter acknowledged 'that broader issues facing the Australian Government in the area need to be resolved before a negotiated boundary is likely to be agreed'.

As the Australian government was working through how to respond to the oil companies, NSW state Liberal MP John Dowd released a report based on his interviews with Timorese refugees in Portugal. Dowd's report for the International Commission of Jurists found there had been a clear breach of international law when Indonesian troops invaded Dili in 1975. Dowd's report was particularly problematic for the government because he was one of them – a lawyer and a member of the NSW parliament representing the governing Liberal Party. His report, however, was treated with the same disregard as Dunn's. The embassy's assessment in a handwritten note read: 'There's not much substance in this.'[48]

Despite the best efforts of the DFA, negative stories about Indonesia's occupation of East Timor continued to be published in the Australian media. In this environment Woolcott accepted an offer from respected journalist Richard Carleton to balance up Indonesia's 'bad press' on East Timor. Woolcott persuaded his Indonesian contacts to give Carleton a visa. Carleton visited Indonesia in July with an Indonesian film crew. Following the visit he wrote to 'Dick' to thank him for all his 'help and assistance'. His letter said the conclusion reached in the film was that the Indonesian takeover was brutal. Carleton described his interview with two priests, who showed him the 'window from which they watched more than 150 people shot on Dili wharf', and then detailed other allegations of massacres, rape and looting and reports that Fretilin was being 'successfully starved out'.[49]

In a handwritten note to his embassy colleagues, Woolcott observed that 'some of this does not look good'. I doubt he was referring to the substance of Carleton's report.

Carleton's visit generated headlines: 'Indonesians shot 150 on Dili pier', 'Inside the island of agony', and 'The place of death in Timor' about the Balibo journalists.[50]

In the articles Carleton described how he managed to give the slip to the 'dozen escorts, guards, interpreters and hangers-on that the Indonesian Government had provided'. He detailed eye-witness accounts of the Dili wharf massacre and estimated 50,000 Timorese had died in the invasion and decolonisation process, which 'could be the bloodiest on a per-capita basis in modern history'. His account corroborated the earlier reports by Dunn and Dowd, which perhaps explains why there was only minimal commentary concerning Carleton's visit in the Australian archives. However, on a visit to the British National Archives I found a Foreign Office file that noted:

> [It] was not only their content, but the fact that Carleton wrote these articles which has astonished and upset the Australian Embassy. Apparently at no

time did Carleton give any indication that he
intended writing for the press about his visit. The
Indonesians were also considerably upset and let the
Australian Ambassador know this.[51]

Carleton's articles prompted Liberal MP Michael Hodgman to lodge a protest note signed by eighty parliamentarians with the Indonesian Embassy in Canberra. But the articles had less impact on Fraser's and the DFA's actions. At a post-ASEAN summit just days after the first reports of Carleton's visit were published, Fraser told Suharto that 'he would like to see Timor buried as an issue between the two countries as soon as this could be done in Australia'. And two days after Carlton's stories appeared in the *Age* the DFA advised Peacock that

> the exploration permit problem and, indeed the likely
> need in the near future to negotiate with Indonesia on
> de-limitation of our respective 200 miles zones of
> maritime jurisdiction, will require the Government
> to give fairly early consideration to the East Timor
> 'recognition' question.

Woolcott sent another cable on 16 August suggesting the government state that it was necessary to deal 'with Indonesia as the administering authority if we are to effect reunions or extend further humanitarian aid to those in East Timor and it might assist these causes if we were now to acknowledge this unequivocally' – the humanitarian spin.

Woolcott was finally getting traction in Canberra. A colleague from the Department of Prime Minister and Cabinet phoned him on 5 September 1977, to let him know 'privately' that the 'political will at the highest level is quite decided' on the need for the 'helpful shift in our Timor policy' the Indonesians were expecting, and that it was 'just a question of the moment'.[52]

Carleton's twenty-minute documentary on the invasion of East Timor focusing on the deaths of the Australian journalists in Balibo aired on the ABC on 21 October. A week later another caller from Canberra told Woolcott that while the foreign minister wanted 'to move this year', the announcement of the federal election on 10 December 1977 had 'intervened'.[53]

Nevertheless, Australia was slowly edging its way toward recognition. On 28 November, when East Timor came before the UN General Assembly, Australia again abstained from a vote that 'reaffirmed the inalienable right of the People of East Timor to self-determination and independence and the legitimacy of their struggle to achieve that right'.

Woolcott met with Murdani and Minister of State Sudharmono in early December. Murdani told him that 'a search and destroy operation against Fretilin remnants had been under way since late October' involving the use of 'aircraft and selective bombing' and that 'Indonesia had decided to "stop pussy footing around and go in after them"'. According to Woolcott, Murdani said that

> the previous policy of seeking to deny residual Fretilin forces food and supplies and of waiting for them to surrender and accept the Presidential amnesty had now been replaced by a policy of seeking to put an early end to Fretilin residual resistance.[54]

Murdani said there could be as 'many as 200,000 people in the "no-man's land" area who might be attracted by handouts of free food'. This was a third of the population – surely an indication that Indonesia was not just 'starving out' the 'Fretilin remnants', but the entire population.

Sudharmono told Woolcott that Suharto's planned visit to Australia was conditional – the 'Timor question would have to be "settled" before Suharto would visit Australia'. Woolcott surmised that 'settled' meant 'de facto recognition by the government or at the least some clear acknowledgement by the government of its acceptance of the fact of incorporation'.

Three days later, on 10 December 1977, the Fraser government won the election.

Woolcott's telephone informants had been correct. Despite continuing domestic opposition to the invasion and Fretilin's ongoing resistance, following the election Fraser moved to resolve the recognition and seabed boundary issues.

CHAPTER 5
THE PRICE OF RECOGNITION

As the Fraser government was settling into its second term, the Indonesian military embarked on the campaign of 'encirclement and annihilation' that Murdani had briefed Woolcott about in December. Murdani also briefed the US State Department, advising them in early January that 'resolving the problem by military offensive would take at least until April or July of this year. Trying to outwait the guerrillas or starve them out could take at least a year longer.'[1] Murdani referred to the 'several hundreds of thousands' in remote areas not controlled by Indonesia or Fretilin. The US cable then notes that 'Australia's Ambassador Woolcott, who follows developments in East Timor closely, generally agreed with these figures but put the number of people in "no-man's land" up to possibly half a million.' This was five-sixths of the population.

The despairing reality of life in East Timor is captured

in a letter from a priest published in the *Nation Review* in January 1978:

> The war continues with the same fury as it had started ... The bombers did not stop all day. Hundreds of human beings died every day. The bodies of the victims become food for carnivorous birds (if we don't die of the war, we will die of the plague), villages have been completely destroyed, some tribes (sucos) decimated ... and the war enters its third year with no promise of an early end in sight. The barbarities ... the cruelties, the pillaging, the unqualified destruction of Timor, the executions without reason, in a word all the 'organised' evil, have spread deep roots in Timor. There is complete insecurity and the terror of arbitrary imprisonment is our daily bread (I am on the persona non grata list and any day I could disappear). Fretilin soldiers who give themselves up are disposed of – for them there is no prison. Genocide will come soon, perhaps by next December.

A week after this story was published, on 20 January 1978, the Australian government announced that it had decided to accept East Timor as part of Indonesia.

The explanation in Peacock's media release was that Australia needed to deal directly with the Indonesian government 'as the Authority in effective control' in order to progress family reunions and the rehabilitation of Timor.

The final line of the media release, which is one of the last documents in the Downer Compilation, states that although the Australian government 'remains critical of the means by which integration was brought about it would be unrealistic to continue to refuse to recognise de facto that East Timor is part of Indonesia'.

The DFA's official history of Australia's engagement with South East Asia, *Facing North*, states that 'the basis for the government's position was that Indonesian control was effective and covered all major administrative centres'.[2] This links back to one of the legal requirements for de facto recognition – effective control of the territory.

THE FRASER GOVERNMENT'S SECOND ATTEMPT AT NEGOTIATIONS

Peacock's submission to Cabinet that led to the announcement had bypassed the usual processes in an attempt to prevent leaks and the revival of what the DFA called the anti-Indonesia campaign. Only a small number of DFA officials were aware of the submission.[3]

The heavily redacted submission shows that he recommended avoiding the de facto/de jure issue by simply stating that 'Australia should no longer sustain its public objection to the integration of East Timor into Indonesia.'[4] Peacock also recommended that Australia advise Indonesia that it was prepared to treat East Timor as part of Indonesia for the purpose of its development aid programs, and that 'appropriate Departments prepare for Ministerial consideration Australia's position for negotiations to complete the demarcation of the sea-bed boundary':

> In response to any Indonesian query about the seabed boundary in the Timor Sea the Government indicate that it is prepared to begin negotiations to complete the demarcation of the sea-bed boundary between Australia and Indonesia as soon as is mutually convenient.

The recognition decision was clearly intended to allow Australia to realise its long-held ambition to close the Timor gap. And it would fulfil Indonesia's long-held objective of securing Australia's recognition of Indonesia's sovereignty in East Timor.

But there was a hitch. The minute detailing Cabinet's decision is fully declassified. It shows that Peacock's recommendation concerning the commencement of seabed boundary discussions with Indonesia was endorsed. However, Cabinet's decision differs in one significant respect. Peacock's submission recommended avoiding using the word 'recognises' because of 'certain international legal considerations', and instead using 'phrases such as "full", "formal" or "definitive" acceptance'. The Cabinet decision states, 'that the Government recognises, de facto, that East Timor is part of Indonesia.'

The use of 'recognises' and 'de facto' against Peacock's advice completely undermined Peacock's strategy to 'slip' into de jure recognition.

There is a draft media release attached to the submission. The version released by Peacock on 20 January is exactly the same except for the insertion of 'de facto' in the last line. Neither version mentions the start of Timor Sea seabed boundary negotiations.

I was still digesting the implications of all this when I read the background paper attached to Peacock's Cabinet submission. It notes that just before Fraser's visit to Indonesia in October 1976 'serious consideration was given to the possibility of Australia according some form of recognition to incorporation, but the idea was rejected', because of 'timing' issues. The paper then describes how, following Fraser's visit, the government closed illegal radios in Darwin used to contact Fretilin, 'frustrated suspected illegal visits' to East Timor, and generally acted 'as if East Timor were part of Indonesia'. It continues:

> At that stage it seemed there would be a natural and steady progression to some form of de facto recognition *[five lines of text are redacted]*. This progression was, for all practical purposes, halted by the activities and reports of observers such as Mr Jim Dunn, Mr John Dowd and, to a slightly lesser extent, Mr Richard Carleton, stimulating, as they did,

> continuing criticisms of Indonesian policies in the press, by several Members of Parliament and a few vocal critics elsewhere in the community.

There is no discussion of the substance of the 'activities and reports' of Dunn, Dowd and Carleton in the declassified text, of the 200,000 deaths, the rape, torture and starvation in East Timor. Perhaps it is redacted but I doubt it. I suspect that most of the redactions concern the seabed issue. While the Cabinet submission specifically recommends starting negotiations with Indonesia to close the Timor gap, there is no discussion of this recommendation in the declassified text.[5] Nor is there an explanation for keeping the decision to start maritime boundary negotiations with Indonesia out of the media release. Or an explanation for why Cabinet decided to insert the qualifying phrase 'de facto'. The only way of knowing why is for the Australian government to fully declassify a forty-year-old Cabinet submission.

The government's decision to give de facto recognition to the Indonesian occupation was condemned by the ALP, the Catholic Commission for Justice and Peace, the Australian Journalists' Association, the Australian Council for Overseas Aid and a range of East Timor solidarity groups. Two members of the government broke ranks and criticised the decision.

The insertion of 'de facto' in the media release led to another year of legal gymnastics and political contortions.

Peacock was later advised, 'For Australia to claim, while negotiating on incidents of sovereignty to territory, that our negotiating partner was not legally entitled to exercise sovereignty over that territory would involve a degree of unreality.'[6]

Woolcott downplayed the significance of the problematic phrase in Peacock's media statement, assuring Mochtar that 'the Australian Government has decided to accept East Timor as part of Indonesia'.[7] He told Canberra that Mochtar said, 'The important thing was that the way was now open to strengthen further our bilateral relations and to deal with practical matters like an extradition treaty and the completion of the seabed boundary'.[8]

In February 1978 the DFA was forced to prevaricate with an 'in due course' response when Indonesia sought to set a date for the start of seabed talks. Mochtar turned up the pressure, repeating that 'all that was needed to close the gap was a ruler'.[9]

The link between Australia's recognition and Timor Sea negotiations with Indonesia remained secret until the *Australian Financial Review* published a story in February stating that the 'Federal Government's recognition of East Timor as Indonesian territory has paved the way for Australian participation in a multi-million-dollar oil exploration program.'[10] The article, clearly sourced from oil industry figures, referred to drilling 'on an area of the seabed the ownership of which was long disputed by Australia and East Timor's former colonial rulers, Portugal':

> Now that Indonesia's takeover has been recognised, oil companies are expecting early negotiations between Canberra and Jakarta to finalise the mid-ocean border line. Had the Portuguese had their way, the border would have been drawn straight across the potential oil field. But oil company officials said yesterday they were now expecting an agreement far more favourable to Australia.

The article explained there were two main companies planning to explore the region: Woodside and the French government–owned Aquitaine-Elf that had the permit over the highly prospective Kelp area in the middle of the gap, overlapped by Oceanic. Aquitaine's Australian exploration manager told the *Australian* that the federal government had long been aware that no company would explore the area until the border dispute was settled:

> No one would want to find oil there without knowing who owns it. But we are not expecting any major problems over the border now because of the border lines already agreed to by Indonesia on either side of the disputed area. If these two lines are just joined together, there will be no trouble at all.

The following day in parliament, Labor senator Cyril Primmer said that if 'even part of that news item is true, it means clearly that the decision to recognise integration was taken in order to settle the dispute about the seabed border between Australia and East Timor'. Primmer accused Peacock of deliberately delaying the announcement until after the federal election because the government was aware of the depth of public support for East Timor. He also challenged claims that Indonesia had 'effective control' in East Timor:

> Only a small part of East Timor is in Indonesian hands; Fretilin is still strong and inflicting heavy casualties upon the Indonesians; in Indonesian-held areas the people are held in virtual slavery; churches, buildings and villages have been destroyed; executions and murder of the East Timorese are going on in the occupied areas.

Primmer concluded that 'surely de facto recognition is completely immoral in these circumstances'.

In response Peacock said Indonesian control was 'effective and covers all major administrative centres' – a statement not borne out by Woolcott's estimates of 200,000 to 500,000 Timorese living in no-man's land. Peacock referred to the need for recognition to enable family reunions and aid delivery (the humanitarian spin) and then said: 'Any other matters to be negotiated are not, were not, and have never been salient considerations at all.'

Peacock's statement to parliament is inconsistent with his declassified Cabinet submission in the archives, which reveals that Australia's decision to recognise Indonesia's sovereignty in East Timor was directly linked with the government's intention to start seabed negotiations.

*

Woolcott ended his posting in Jakarta on 30 March 1978, having failed to achieve the closure of the Timor gap. He was out of favour with Fraser and shifted sideways to become ambassador to the Philippines. His replacement in Indonesia was Tom Critchley, Australia's former high commissioner to Papua New Guinea.

Critchley picked up where Woolcott left off. Reports from Indonesia's most senior generals of a growing humanitarian crisis in East Timor were subject to the same pattern of denial as under Woolcott. Brigadier-General Adenan's information that Fretilin was 'now suffering lack of both ammunition and food' and that some recently 'surrendered' Fretilin supporters 'could not even stand' was met with equanimity by Ambassador Critchley and David Irvine, the embassy's first secretary (and later head of ASIS, then ASIO), and duly passed down the line to Canberra in the same cable that reported Adenan's comment that 'Mochtar was very anxious to commence negotiations on the Seabed Boundary between Indonesia and Australia.'[11]

The record of Critchley's meeting with Indonesian General Yusuf is equally disturbing. Yusuf told the ambassador that he had 'just returned from East Timor', where 'one of the biggest problems was the 270,000 women and children to care for'. Critchley commented that the figure seemed 'unduly high as the total population was only between 500,000 and 600,000'. General Yusuf did not revise his estimate and, according to Critchley, gave the 'impression that he was well briefed and in command of his subject – East Timor'.[12] This information does not appear to have affected Critchley's assessment of Australia's relationship with Indonesia.

Back in parliament, the ALP Opposition asked if it would be necessary for 'Australia to change recognition of Indonesia's annexation of Timor from de facto to de jure in order to facilitate the negotiations over the exploitation of seabed resources in that area'.

The Legal and Treaties Branch of the DFA advised Peacock, in a brief out of a *Yes Minister* script, that if the government didn't want seabed negotiations with Indonesia concerning waters off East Timor to give rise to a situation of de jure recognition,

> it might be necessary to affirm at the relevant time that the negotiations do not affect Australia's policy on recognition. The difficulty with this option, however, is that such a statement would appear to others, including the Indonesian government, to be in contradiction with our behaviour.[13]

This brief also resurrected Parsons' advice to Peacock from August 1976 about 'slipping' into recognition, concluding that: 'In all the circumstances, the Government may prefer to "slip" into de jure recognition of Indonesia's incorporation of East Timor.'[14]

This time there is a tick next to this advice.

The DFA again moved quickly. On 27 July Peacock announced Australia's decision to negotiate maritime boundaries with its neighbours, New Zealand, the Solomon Islands and France. Peacock said the negotiations were 'not expected to be complicated', as in each case Australia would 'seek to negotiate agreements using the median line'.[15] Needless to say, Australia's maritime negotiations with these countries were not complicated by billions of dollars worth of oil and gas.

Peacock didn't announce that in the same meeting the Australian Cabinet had also decided, in relation to Indonesia, that 'the Timor gap should be closed,' which amounted to a decision to give de jure recognition of Indonesia's sovereignty in East Timor.[16] This remained Australia's secret agenda.

REPORTS OF MASS STARVATION IGNORED

The embassy in Jakarta continued to be briefed on Indonesia's military operations in East Timor, including a new initiative 'involving 29,000 Indonesian troops' that were going to move from 'west to east forcing rebels toward eastern end of Timor'.[17] Again, the implications of this initiative for the men, women and children of East Timor went unregistered.

Many of the files concerning Indonesia's military operation in this period are digitised as a result of Clinton Fernandes' research into what Australian officials knew about Indonesia's deliberate mass starvation campaign. Two briefs stand out. Both involve Douglas Campbell, who was in charge of aid programs at the Australian embassy in Jakarta.

In mid-August, Campbell was sent to West Timor with a colleague from the embassy, Patrick Alexander, to report on the aid situation. They reported they were shown photographs of the condition of the refugees in Bobonaro, a village in East Timor, that 'depicted many sick, starving and malnourished women and children, typical of famine scenes throughout the world'.[18] A Dutch priest who had been across the border into East Timor told them that the 'Indonesian authorities were "doing almost nothing" to alleviate the situation … He estimated that at least 1.5% of the refugees were dying monthly and that in some groups the death rate was around 8% per month.' The priest said Fretilin controlled large areas of the countryside. Campbell and Alexander reported that the situation in East Timor had grown worse over the last twelve months and that, unless there was 'firm Government action, people will continue to die of hunger and the Government will find that winning the hearts and minds of the people will be made more difficult by its own apparent lack of concern'.

A lack of concern echoed in Jakarta and Canberra.

As Campbell told me in an interview in 2014, 'No one thought about the welfare of the East Timorese people.'[19]

*

In Canberra, the permit holders were keeping up the pressure. Lawyers for two unidentified companies 'exploring for oil in the Timor Sea' wrote to Peacock on 4 September seeking to have a 'boundary fixed before commencement of drilling'. The letter stated:

> You will appreciate that if a well is drilled and oil discovered in an area near to the northern boundary of the Permit, it may affect a decision on the part of the Indonesian Government as to the proper location of the boundary.[20]

The lawyers sought an assurance that the government would 'ensure delineation of the boundary in such a way as to protect existing Permitees'. A few days later, Ambassador Critchley, accompanied by a select group of journalists and nine other ambassadors, including from the US, Canada and New Zealand, visited East Timor. They were escorted by Mochtar, who was now Indonesia's foreign minister. The London *Sunday Times* reported that the trip was designed to 'achieve world acceptance' of Indonesia's claim to sovereignty in East Timor.[21]

As Parsons had flagged in his August 1976 brief to Peacock, 'an official visit to East Timor by the Ambassador' would be a way to 'slip into a form of recognition without having to say so publicly'.[22]

In Remexio, in the mountains behind Dili, the delegation visited a 'transit camp' and witnessed appalling conditions and mass starvation. Melbourne *Herald* journalist Richard Gill described naked children, with 'the piercing gaze, emaciated limbs and distended bellies of advanced malnutrition'.[23] They heard reports of tens of thousands of refugees 'coming out of the hills' starving after 'three years away from civilization', wrote David Jenkins from the *Far Eastern Economic Review*.[24] On the front page of the *Sydney Morning Herald* Peter Hastings described thousands of refugees pouring into relief centres from the hills, suffering from malnutrition and malaria.[25]

None of the reports blamed the Indonesian government for the appalling condition of the Timorese. Hastings attributed their condition to Fretilin. According to Jenkins, Indonesia was 'saddled with this enormous humanitarian and development problem as a result of its takeover of East Timor in 1975', implying the situation was inherited from the Portuguese. Gill quotes an ambassador's observation that the 'Portuguese have a lot to answer for. And seeing these people makes a mockery of any suggestion for a United Nations referendum.'

The DFA was aware of the Indonesian tactic of 'starving' out the Fretilin resistance.[26] In addition to the direct briefings the diplomats at the Australian embassy received from Murdani, they also had access to intelligence reports from Australia's agencies.[27] Major international newspapers later reported that 'starvation was a deliberate policy to crush opposition to Jakarta's annexation'.[28] Here was the tragic consequence of that policy – dying children, hundreds of thousands of displaced and desperate people – and yet it would appear from the Australian media reports that no one from the DFA corrected the journalists' false impression that the condition of the Timorese had nothing to do with Indonesian policies or administration.

The DFA appears to have also gone along with Mochtar's spin that for aid to flow, donor countries needed to acknowledge Indonesia's sovereignty. Both Hastings and Gill make this point, Gill most explicitly. He wrote that Mochtar

> tells me it might be possible for other countries to help East Timor's development with direct government-to-government aid. Dr Mochtar says the only snag could be that donor countries would have to give the aid without strings attached and might also have to specify it is for East Timor when handing it over to Jakarta. This would be an effective admission that East Timor is in fact now part of Indonesia.

The Indonesia/DFA messaging in the media reports is as obvious as it is shocking. As US academics Arnold Kohen and Roberta Quance note, it was a 'crude attempt' by the Indonesian government 'to entice recognition of its claim to East Timor by making that a prerequisite for humanitarian relief assistance'.[29] The only reason aid could not flow to East Timor was because Indonesia would not allow it. Again, the DFA did nothing to correct the record. The files in the archives reveal that as early as Fraser's visit in October 1976, Australia attempted to explain its moves to recognise Indonesia's occupation of East Timor by reference to humanitarian motives, when it was actually the legal consequence of maritime negotiations that created the imperative.

In response to the newspaper reports, Liberal MP Michael Hodgman raised the link between the famine and Indonesia's deliberate use of defoliants to destroy crops in parliament. Frustrated with the lack of Australian government action to address the humanitarian crisis in East Timor, seventy-six Australian MPs from the government and Opposition petitioned the United Nations to allow the International Red Cross and a UN relief mission into East Timor.

THE FRASER GOVERNMENT'S THIRD ATTEMPT AT NEGOTIATIONS

On 18 July 1978 Woodside again wrote to the government,

this time formally seeking a deferral of drilling obligations in the Greater Sunrise area because of 'uncertainty of the seabed boundary'.[30]

Indonesia also kept up the pressure. At the Law of the Sea Conference in Geneva a senior Indonesian official told an Australian delegate that 'the time seemed to be ripe to commence delimitation talks' and that 'Indonesia would generally be content with merely joining points A16 and A17 with a single straight line'.[31]

The lure of a straight-line boundary to close the Timor gap hooked Australia. On 5 October Peacock met with Mochtar in New York. The ministers agreed to open seabed negotiations and to make a public announcement when Mochtar visited Australia before the end of the year.[32] In a Cabinet submission on 27 October 1978 Peacock concludes:

> The uncertainty faced by exploration permit holders in the Timor Sea area, the need for Australia to take action to avoid the lack of negotiation becoming a further irritant in the Australian/Indonesian relationship, and the administrative need for Australia to delimit its prospective 200n.m. fishing zone provide pressing reasons for us to start delimitation negotiations with Indonesia. For these reasons, I consider that substantive negotiations with Indonesia should commence this year and that by such negotiations the Government acknowledge that Australia has moved to de jure recognition.[33]

*

In East Timor, the carnage continued. Resistance leader Xanana Gusmão recalls being part of a group of 140,000 Timorese encircled by Indonesian troops on the east of the island at Mt Matebian in November 1978. They were 'bombed mercilessly with napalm and scatter bombs by Indonesian forces twenty-four hours a day for weeks on end'.[34]

In a token response to the humanitarian crisis, Peacock approved a $250,000 grant for the Indonesian Red Cross for East Timor.

In mid-November, Douglas Campbell became the first Australian diplomat to make a solo visit to East Timor since the Indonesian invasion. He was sent to Dili to supervise the arrival of an Australian shipment of nutritional biscuits and skim milk powder. Campbell's report, sent to the DFA and various intelligences agencies in Canberra, said that according to the Indonesian Red Cross's own figures, there were over 200,000 refugees in East Timor, or 30 per cent of the population. Most were women and children and 'almost all were suffering from malnutrition and many from malaria. The death rate was high and extremely high among the children.' Refugees were housed in camps for three to four months and then 'released' in groups of around 300. He also reported that the Indonesian Red Cross was unable to answer his basic questions about the assessment of need, death rates or the administration of the camps.[35]

In an interview in 2014 Campbell recalled that his report went to Ambassador Critchley.

> Even though it's a catalogue of terrible things – nobody said a word to me. We had weekly senior officer meetings – not raised. I would not have raised it in front of people because there were trade and cultural personnel at these meetings too. I would have only raised it if Critchley or another had brought it up. The only thing I heard was that the Indonesians were not pleased with me as I had spoken to someone in East Timor who was persona non-grata. The Indonesian security people complained ... Australia was desperate to be seen to be putting more aid into East Timor, but it was pointless when the Indonesians weren't letting the aid go through. They were feeding the army and that's it.[36]

Campbell was told by a colleague in Canberra that the department had been able to 'make good use' of his report. He later discovered the only parts used were 'one or two bits which said positive things ... Before we went there were caveats from Ambassador. He referred to Kissingerian realism ... You had to be realistic and pragmatic.'

Campbell's report was circulated in Canberra. The following day, 15 November, Cabinet endorsed Peacock's submission and agreed to announce de jure recognition and commence negotiations with Indonesia to close the Timor gap.[37]

This explains why Australia became the only Western nation to give de jure recognition to Indonesia's illegal occupation of East Timor – it was an unavoidable legal consequence of negotiating with Indonesia to close the Timor gap.

*

In East Timor on 22 November, Gusmão was forced to give up the defence of Mt Matebian and tell the population to surrender. The Fretilin meeting place was bombarded. Gusmão recalls:

> A deafening explosion had been just a few metres ahead of us. Pieces of burnt, dirty flesh were plastered to the stones, the trunks of trees embedded with sharp edged metal that had dismembered the bodies of Companheiros. I, who used to yell for others to find cover while looking after other fighters, now suffered a terrible fear of aerial attacks.[38]

A week later Peacock advised the Australian delegation in New York for the first time to vote *against* the UN resolution supporting self-determination for East Timor.

Mochtar visited Australia in mid-December. At a joint media conference Peacock said it would be obvious that:

the negotiations, when they commence, of seabed boundaries adjacent to East Timor will signify a de jure recognition by Australia of the incorporation of East Timor into Indonesia. That is a matter of law, but we also have a duty to be closing that gap.[39]

On 31 December 1978 the leader of the Timorese resistance, Nicolau Lobato, was killed following a six-hour battle. His mutilated body was flown to Dili and presented to the minister for defence, General Jusuf.[40] A body said to be Lobato's was shown on television throughout Indonesia.[41] Soon after, Xanana Gusmão became the leader of a desperate and devastated resistance.

NEGOTIATIONS BEGIN

After three years of diplomatic manoeuvring and the death of an estimated 200,000 Timorese, the first round of seabed negotiations between Australia and Indonesia to close the Timor gap commenced in February 1979. No Timorese were at the table. Australia was negotiating with a state that, according to the UN, was an illegal occupier.

There were four issues left to be negotiated not covered by the 1971 and 1972 treaties; the Timor gap between Australia and East Timor; the seabed west of the 1972 boundary referred to as the 'western sector'; the seabed boundary between Christmas Island and Java; and the final delimitation of the water column along the whole area between Australia and Indonesia. A Cabinet submission identified the 'critical issue' for Australia as the seabed boundary in the Timor Gap, where 'Australia has little room to manoeuvre' if it was 'to ensure a share in the Kelp prospect, potentially a large petroleum resource straddling the straight-line closure'.[42]

Instead of closing the Timor gap with a straight line, as the DFA negotiators hoped, Mochtar argued strongly in favour of the median line. He laid out what was essentially the buried 1970 advice from the Australian Bureau of Mineral Resources, arguing there was a common shelf in the Timor Sea and that the Timor Trough was just a seabed 'irregularity'. Later he conceded the geology of the Timor Sea seabed was irrelevant, as the Law of the Sea Conference had agreed on EEZ rights that would result in a median line in the Timor Sea.

Australia continued to insist on continental shelf rights. The talks failed.

Ahead of the next round of talks in May 1979, the Department of National Development provided updated advice on the petroleum prospectivity of the Timor Sea.[43] The report said the 'Kelp prospect' was likely to match the potential of the Greater Sunrise, and valued the 'so far discovered reserves of the Troubadour dome' at $1 billion (the equivalent of approximately $4.6 billion in 2017 and a massive underestimate). Indonesia refused to budge from the median line at the May talks.

*

In East Timor, between March and September 1979 at least 600 people were executed or disappeared.[44] And the deliberately induced famine continued to decimate the population.

Pictures of the devastating reality of the famine in East Timor were published in the Australian media in November.[45] The photos were taken by Peter Rodgers, who had left the Jakarta embassy in September 1977 to become the Fairfax correspondent in Indonesia. He later rejoined the DFA. Rodgers visited East Timor with the approval of the Indonesian military to report on aid arriving from Catholic Relief Services and the International Committee of the Red Cross, which had finally been allowed into the territory. His story featured photos of starving women and children, and said that the population had declined by about 100,000 in four years.

Here was a former senior DFA diplomat reporting a 100,000 death toll. Yet, like the journalists covering the ambassadors' September 1978 tour, Rodgers attributes the appalling condition of the Timorese not to 'deliberate intent on Indonesia's part' but to 'violence and neglect'.[46]

The famine continued, and so did the orchestrated violence.

In 1981 Indonesia commenced the 'Fence of legs' offensive which involved 60,000 Timorese men, young and old, forced to march in human chains, one starting in the west and one in the east. They were marched across the island with the Indonesian military at their backs,

capturing and executing anyone suspected of supporting Fretilin or its military wing, Falintil. Many died of starvation or exhaustion or were executed by Indonesian troops for allowing Fretilin supporters to slip through. The marches ended at Lacluta in September, where hundreds of men, women and children were massacred after surrendering to Indonesian troops.[47]

In Canberra, the pressure from the mining companies and their financial backers continued. Indonesia and

Australia were no closer to agreement at the conclusion of the fourth round of talks in October 1981.

The Law of the Sea Conference finally settled the text of the United Nations Convention on the Law of the Sea in 1982. More than 150 countries had participated in the drafting of the convention over fourteen years. Australia and Indonesia were both set to gain enormous new maritime areas as a result of the 200-nautical-mile EEZ right giving sovereignty over the seabed and the water column.

However, in the Timor Sea, where Australia and Indonesia's EEZs overlapped, the convention bolstered Indonesia's claim to the median line.

THE HAWKE GOVERNMENT CARVES UP THE TIMOR GAP

In opposition the ALP strongly supported East Timorese self-determination, and at the 1982 ALP National Conference adopted a policy calling for the suspension of all defence aid to Indonesia until it withdrew from East Timor.

At the March 1983 federal election, the ALP returned to government under the leadership of Bob Hawke. The ALP policy meant Australia's new government did not recognise Indonesia's sovereignty in East Timor. Negotiations with Indonesia to close the Timor gap were again off the table.

It didn't take long for the Hawke government to change its policy.

A 200-nautical-mile EEZ was in the final draft of the 1982 Law of the Sea Conference. The continental shelf concept survived, but it was only relevant where a continental shelf extended beyond 200 nautical miles. This meant a median line in the Timor Sea, unless Australia and Indonesia agreed otherwise.

On 29 March 1983 Cabinet authorised Foreign Minister Bill Hayden to publicly state that the government 'notes that Indonesia has incorporated East Timor into the Republic of Indonesia', which Hayden did during a visit to Jakarta in April.[48]

Following the DFA script, the Hawke government sent a parliamentary fact-finding mission to East Timor in July 1983. Four months earlier, the East Timorese had signed a ceasefire with the Indonesian military. The ceasefire was shaky but held until after the visit by the Australian parliamentary delegation led by Bill Morrison, who had been the ALP's defence minister under Whitlam in 1975. James Dunn writes it was soon clear that Morrison had little interest in uncovering human rights abuses or hearing the Timorese perspective on integration and the ceasefire. Gusmão and Ramos-Horta later accused Morrison of betraying a group of guerrillas who tried to set up a meeting by tipping off the Indonesian military.[49]

Once the Australian delegation left East Timor, the ceasefire collapsed. The report of the Commission for Reception, Truth and Reconciliation in East Timor, known as the *Chega* report, found that Indonesian forces executed more than 200 civilians, mostly men, in the months of September and October 1983.[50] The report also found that many of the surviving women of Kraras were forced into sexual slavery.

On script, Hawke then framed the need to recognise Indonesia's sovereignty as a humanitarian issue. The government suggested that those seeking to support East Timor would 'do better to concentrate on helping the province through provision of development aid and assistance'.[51]

Yet the archives reveal that the Hawke government's primary motivation for recognition was again the government's desire to negotiate with Indonesia to close the Timor gap.

On 16 August 1983, Cabinet decided that Australia should 'continue its policy of negotiating maritime agreement with neighbours'.[52] Talks with Indonesia were scheduled for early February 1984. Cabinet was advised that the recent discovery of oil by BHP at the Jabiru field on the North West Shelf had renewed interest in the status of Australia's boundaries with Indonesia.

Hawke and Hayden pushed through a change of wording in the ALP policy that was woolly enough for Hayden and Mochtar to immediately agree to continue negotiations.

Soon after, Australia recognised Indonesia's sovereignty in East Timor. And Indonesia agreed to a 'temporary resource sharing arrangement' with Australia as allowed under the Law of the Sea Convention, an arrangement that preserved Australia's interest in oil-rich areas north of the median line.[53]

Hawke chose Indonesia's National Day, 17 August 1985, to state that Australia recognised 'the sovereign authority of Indonesia'.

Xanana Gusmão listened to Hawke's announcement on Radio Australia.[54] In a recorded message to supporters Gusmão referred to the ALP's 'pragmatic' change of policy and linked it to the 'extremely important fact of the current negotiations concerning the oil explorations of the Timor Gap'. He said, 'Bob Hawke's hands are stained with the blood of the East Timorese.'[55]

Gusmão's message had no obvious impact in Canberra.

The minister for resources and energy, Senator Gareth Evans, announced a new round of negotiations, saying he expected the talks would 'concentrate on the identification of boundary lines for a joint development zone in the vicinity of the Timor Gap, and the development of an appropriate regime for exploration and development of petroleum resources within it'. Australia's key objective was now to ensure that the shared zone was as small as possible, and that the Greater Sunrise field, as well as the highly prospective areas near the western end of the gap, were outside the zone.

In 1986 Indonesia became a signatory to the UN Convention on the Law of the Sea. If Indonesia had insisted

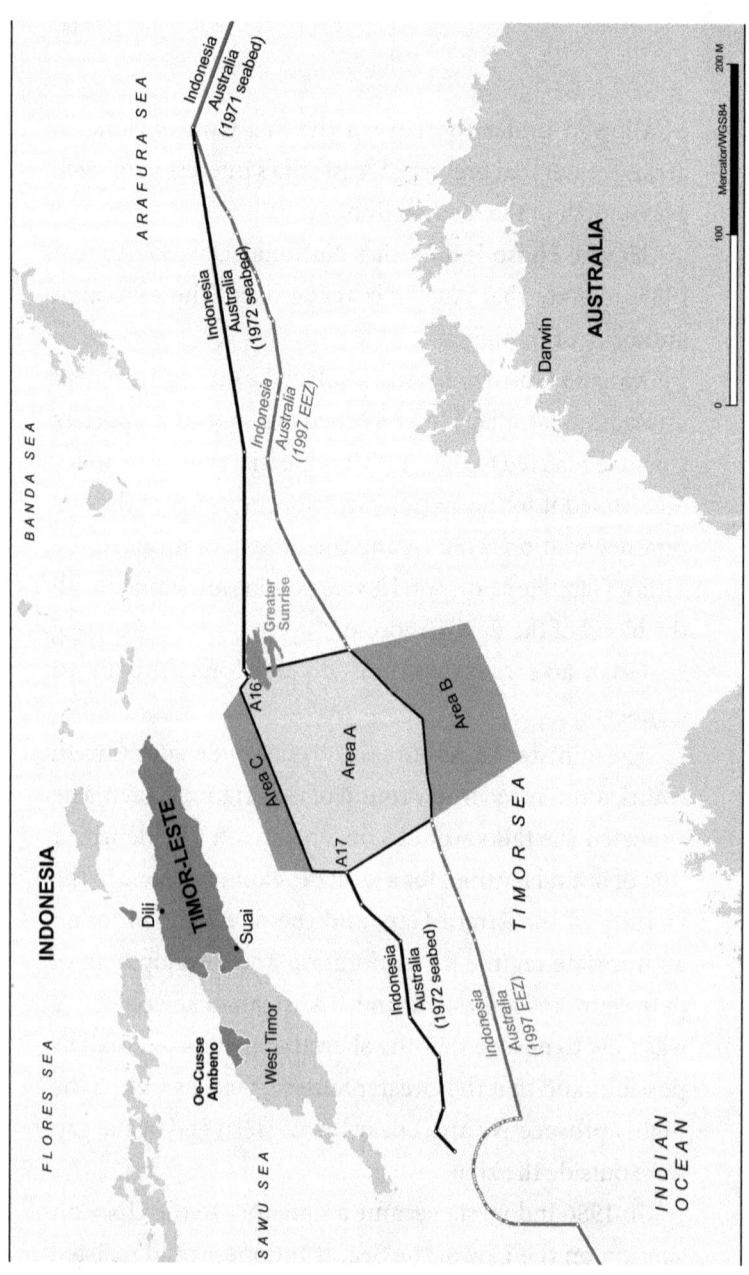

Figure 2: Timor Gap Treaty 1989

Australia agree on a boundary in accordance with EEZ arrangements in the convention, the result would have been a median line and all revenue from fields north of that line would have gone to Indonesia.

THE TIMOR GAP TREATY OF 1989

Three years later Australia and Indonesia agreed to divide the Timor Sea between Australia and East Timor into three 'revenue sharing' zones, in which they would jointly issue exploration and exploitation permits and share revenue according to formulas set out in the Timor Gap Treaty.

Gareth Evans was now foreign minister and Richard Woolcott secretary of the DFA. In December 1989, in one of the most ill-conceived media events in Australian political history, Evans posed with a glass of champagne with Indonesia's foreign minister, Ali Alatas, as they signed the Timor Gap Treaty in a jet over the Timor Sea.

Evans had reason to celebrate. Indonesia had conceded nearly 80 per cent of the Greater Sunrise field to Australia as the eastern lateral boundary pushed inward in Australia's favour. The western boundary also pushed inward, just far enough for three soon-to-be discovered oil and gas fields to be outside the shared zone and in waters claimed by Australia.

Despite having no direct contact with the world outside East Timor, Xanana Gusmão closely followed the

negotiations between Australia and Indonesia to carve up the Timor Sea via newspaper clippings smuggled through resistance networks and international radio services. In a recently uncovered letter to ALP senator Gordon McIntosh in 1988, he describes Australia's plans for joint exploration of the Timor gap as 'stealing' and 'plundering'.[56]

Three years later, when lawyer Robert Domm risked his life to interview Gusmão in his remote mountain camp, Gusmão described the Timor Gap Treaty as illegal, illegitimate and criminal. He said Australia had been an 'accomplice in the genocide perpetrated by the occupation forces'. Its negotiation of the treaty, he said,

> shows the dirty, cynical and criminal policies
> practised by the Australian government in regards to

> East Timor. We feel betrayed that a country with Western values should profit from our people's blood by participating in this rapacious exploitation of something that is in fact legitimately ours.[57]

A month later, as Australia was preparing to sign an agreement with Indonesia to award contracts to Phillips Petroleum, Shell, Woodside and other companies to explore and exploit resources in the Timor Gap Zone of Cooperation, news broke of the massacre of over 200 students peacefully protesting at the Santa Cruz cemetery in Dili. The massacre was filmed by journalist Max Stahl and screened around the world. Foreign Minister Gareth Evans described what had happened in Dili not as an act of the state, but 'the product of aberrant behaviour by a subgroup within the country'. He went ahead and signed the Timor gap production-sharing agreement with Indonesia.

Gusmão was captured by Indonesian forces soon after the Santa Cruz massacre and sentenced to life imprisonment in a show trial in Dili. He continued to lead the Timorese resistance from prison. Imprisonment allowed Gusmão to be more engaged in the diplomatic struggle for independence. He began learning English and a steady flow of smuggled material from Fretilin kept him on top of events in East Timor, Indonesia and at the UN.

*

By the mid-'90s, after decades of diplomatic scheming, it seemed Australia's interests in oil and gas resources north of the median line were secure. They were so secure that in October 1994 Australia signed the UN Law of the Sea Convention, which came into force the following month.

There was good news for Australia in June 1995 when the International Court of Justice handed down a decision on a lawsuit against Australia initiated by Portugal in 1991. Portugal had claimed that the Timor Gap Treaty violated East Timor's right to self-determination and Portugal's rights as the responsible administrative authority. The court upheld East Timor's right to self-determination, but could not invalidate the Timor Gap Treaty because Indonesia had not accepted the court's jurisdiction.

The mid-'90s were also good years for the mining companies with Australian permits. In 1994 Phillips Petroleum discovered the Bayu-Undan gas field 135 miles south of Timor and 270 miles north-west of Darwin. Later that year Woodside and its partners started development of the Laminaria field that was located just to the west of the Zone of Cooperation in so-called Australian waters. In 1995 Woodside discovered the nearby Corallina oilfield, also, fortuitously for Australia, just outside the zone.

THE DOWNER YEARS

The ALP government was defeated in March 1996 and the Liberal/National Party Coalition returned to power under the leadership of Prime Minister John Howard. Alexander Downer became foreign minister, a post he held for the next eleven years. Downer was to become a very active proponent of Australia's Timor Sea oil agenda. Woolcott's son, Peter, who had followed in his father's footsteps into the DFA, was appointed Downer's chief of staff.

In September 1996 Phillips lodged an application with the Northern Territory government to build a massive LNG plant and pipeline to Darwin from the Bayu-Undan field in the Zone of Cooperation. In October the Woodside consortium discovered Buffalo, another field, near Laminaria, also just outside the Zone.

It seemed that, after nearly four decades of ruthless high-level diplomacy, the DFA had finally achieved Australia's Timor Sea oil agenda. The permits Australia had unilaterally issued in the 1960s were secure. Australia was set to reap billions of dollars in royalties and company taxes from the Timor Sea permit holders.

Australia and Indonesia had also finally started negotiating a fisheries boundary. It was based on median line principles in accordance with the Law of the Sea Convention's EEZ principles. Normally under the convention the seabed would be included in the EEZ, but because the convention did not override existing treaties the seabed was excluded from the EEZ in the Timor Sea. This led to the unusual situation where in a band of the Timor Sea and Arafura Sea, between the median line and the 1972 treaty line, Australia had sovereignty over the seabed and Indonesia over the water column.[58]

But Suharto's declining grip on power and the presidency breathed new life into the Timorese campaign for independence. The Timorese resistance received a major fillip in October 1996, when Roman Catholic bishop Carlos Filipe Ximenes Belo and the leader of the Timorese international campaign, José Ramos-Horta, were awarded the Nobel Peace Prize for their work towards reaching a peaceful resolution of East Timor's volatile disputes with occupying Indonesian forces. In July 1997 Suharto allowed President Nelson Mandela to meet with Xanana Gusmão in Jakarta, and the following year the United Nations auspiced negotiations between Portugal and Indonesia about the future of East Timor.

Suharto was forced to resign in May. The new Indonesian president, B.J. Habibie, announced in June in 1998 that Indonesia would consider a special status for East Timor – a senario that threatened to undo decades of scheming by the McMahon, Whitlam, Fraser and Hawke governments.If the territory separated from Indonesia, the treaties Australia had negotiated with Indonesia concerning the area between Australia and East Timor would cease to exist.

CHAPTER 6
AN INDEPENDENT ADVERSARY

Indonesia's decision to allow, after twenty-four years, a vote of self-determination in East Timor has been attributed to a variety of factors, including the fall of Suharto, the Asian financial crisis and the mistaken belief that the Timorese would vote to stay with Indonesia.

Former Australian prime minister John Howard claims Australia was primarily responsible for East Timor's independence. Howard devotes a chapter of his autobiography to this: 'When asked to list the achievements of my prime ministership of which I am most proud, I always include the liberation of East Timor in 1999.' In 2015 Howard wrote, 'Australia's involvement in the liberation of East Timor in 1999 was one of the more noble things our country has done in many years. It directly led to the birth of a very small country whose people remain very grateful for what we did.'[1]

Howard's assessment of the significance of Australia's

role in the liberation of East Timor has been embraced by the DFA. *East Timor in Transition 1998–2000: An Australian Policy Challenge* was published by the DFA in 2001.[2] It was launched by Downer and purportedly published to provide a 'full and balanced account of Australia's response to the extraordinary foreign policy challenge of East Timor'.

While the Downer Compilation, which was being prepared simultaneously, had a veneer of academic respectability, with an expert review panel and involvement of the Opposition, as Clinton Fernandes notes, *East Timor in Transition* was produced by a 'team of departmental officers who had worked on East Timor over the period' so that 'those who had implemented policy were assessing their own performances within the covers of a book they had themselves written, using material they had themselves selected'.[3]

Like the Downer Compilation, *East Timor in Transition* does not comment on the relationship between Australia's interest in oil and gas fields in the Timor Sea and its foreign policy response to the events across the Timor Sea.

There is no discussion of the fact that if East Timor became independent, the 1989 Timor Gap Treaty between Australia and Indonesia dividing up oil and gas resources would cease to exist and Australia would need to negotiate a new treaty with an independent East Timor. Under this scenario an independent East Timor was a serious threat to Australia's access to oil riches – all of which were on the Timorese side of the median line. Australia's interest in oil and gas in the Timor Sea is mentioned once in the 312 pages, in a six-line section titled 'Timor Gap Treaty'. The brief discussion skates over the enormous diplomatic effort required by the DFA and Alexander Downer to keep Australia's claim north of the median line alive.

The Australian government records from 1998–99 are still in the closed period. I have little doubt there are files discussing legal advice that an independent East Timor would have rights to oil and gas resources to the median line in the Timor Sea – and what this meant to the Australian treasury. Australia and Indonesia made US$1.1 million from oil royalties from the Timor gap in 1998 and the figure for 1999 was predicted to rise to US$2.2 million. But that was peanuts compared with the billions in taxes and royalties expected to flow into Australian government coffers when ConocoPhillips' Darwin LNG plant was operational and Woodside's massive Greater Sunrise field developed.

It is not surprising, therefore, that Australian officials were far from enthusiastic about moves at the UN to facilitate negotiations between Portugal and Indonesia about the future of East Timor.

Events moved fast. In 1998 the National Council of Timorese Resistance, the Timorese leadership, was formed and Xanana Gusmão elected president.

In January 1998 the ALP's shadow foreign affairs minister, Laurie Brereton, pushed through a reversal of party policy to again support self-determination for the people of East Timor. The ALP also recommended renegotiating the revenue-sharing arrangements under the Timor Gap Treaty.

Downer visited Indonesia in July and pushed a plan for East Timor's 'autonomy' within Indonesia, to be followed by a review of the status of East Timor at some unspecified point in the future. He said Australia wanted 'Timor to remain a part of Indonesia' as a 'vote on self-determination would only lead to renewed civil war in the territory'.[4]

A key meeting, not discussed in *East Timor in Transition*, occurred in August 1998. Gusmão secretly met with BHP's senior representative in Indonesia. According to petroleum geologist Geoffrey McKee, the meeting was part of a deliberate tactical shift by the Timorese leadership to 'rob the Australian government, editorial writers, and the Timor Gap contractors of reasons for arguing that independence in East Timor would "tear up the Timor Gap Treaty"'.[5] Details of the meeting, attributed to a well-placed diplomatic source, were reported in the *Sydney Morning Herald* on 20 August 1998. According to the report, Gusmão agreed that an 'independent East Timor would honour – during an interim period – the rights awarded to mining companies under the controversial 1989 Timor Gap Treaty'.

Soon after, for the first time, Downer called for Gusmão's release.[6]

In December 1998 Prime Minister Howard wrote to Indonesian president Habibie. He did not recommend independence for East Timor. He proposed a 'compromise political solution' with a 'built-in review mechanism' which 'would allow time to convince the East Timorese of the benefits of autonomy within the Indonesian Republic'. As Clinton Fernandes writes, it is 'a revisionist distortion' to claim this letter is evidence of Howard's support for East Timor's self-determination.

Downer's foreign affairs advisers, Josh Frydenberg and Greg Hunt (both now ministers in the Liberal–National Party Coalition government), wrote an opinion piece in the *Australian* in mid-January 1999 arguing it was Australia's job to encourage opposition groups in both Indonesia and East Timor 'to support a staged process rather than to make unrealistic demands for immediate independence'.[7]

It was the long-retired Richard Woolcott who stepped back into the limelight and spelled out Australia's real concerns. He told the *Australian Financial Review* that a change in the status of East Timor could 'lead to substantial financial implications for the Government if the Timor Gap Treaty, signed in 1989, were to unravel'.[8] Woolcott claimed the ALP's plans to renegotiate revenue-sharing arrangements could have 'major legal and commercial implications'.

President Habibie rejected Downer and Howard's staged autonomy proposal, and on 27 January 1999 announced the East Timorese would be granted a referendum on independence.

From then on, Fernandes says, 'Australia's diplomacy functioned as an obstacle to East Timor's independence. When the Howard government was eventually forced to send in a peacekeeping force, it did so under the pressure of a tidal wave of public outrage.'[9]

In a replay of the cover-ups and denials of Indonesian atrocities in East Timor following the invasion in 1975, the Australian government ignored intelligence reports throughout 1998 and 1999 that the Indonesian military was arming militias in East Timor. As Shadow Minister for Foreign Affairs Laurie Brereton recalls, it was

> a matter of record that Mr Downer accepted Indonesian Foreign Minister Alatas's denials that the Indonesian military were orchestrating militias in East Timor … He did so at a time when the Australian Government knew from its own Defence Intelligence reports that this was a deliberate strategy to sub-contract out violence against pro-independence supporters … It is also a matter of record that the Australian Government actively argued against pressing Jakarta to accept the early deployment of peacekeepers.[10]

International pressure led to Xanana Gusmão being released from jail into house detention in February 1999. The oil companies and Australia argued that in the event of an independent East Timor, the Timor Gap Treaty should continue, with East Timor taking the place of Indonesia. Downer raised the treaty with Gusmão when they met on 25 February.[11] Gusmão's focus was on seeking the Australian government's help to set up a peacekeeping force in East Timor. His plea fell on deaf ears. The following day DFA secretary Ashton Calvert disputed the assessment of Stanley Roth, US assistant secretary of state, that a full-scale peacekeeping operation in East Timor was unavoidable. Roth argued that in the absence of international action to push for peacekeepers the territory would descend into violence. Calvert argued:

> One of the central themes to achieving a resolution was to convince the Timorese that they had to sort themselves out, and to dispel the idea that the UN was going to solve all their problems while they indulged in vendetta and bloodletting.[12]

Just weeks later, on 6 April, about 2000 people gathered in the compound outside the Catholic church in Liquica in East Timor seeking refuge from Indonesian militia targeting independence supporters. At midday, Indonesian troops and militia arrived at the church and demanded the pro-independence village chief be handed over. Tear gas was thrown into the priest's house where families were seeking refuge. Around sixty people were killed as they fled, either shot or hacked to death with machetes.[13] As his biographer Sara Niner notes, Gusmão had 'warned all, who would listen, of the possibility of such events but no one in the international community would respond to his pleas'.[14] Downer refused to confirm the Indonesian army's role in the massacre and then refused to release a report on the incident prepared by Australian diplomats.[15]

It later emerged that the DFA was also refusing to share intelligence about Indonesia's plans for escalating violence in East Timor with the United States and other key allies. Andrew Peacock was now Australia's ambassador in Washington. In early 1998 Australia's Defence Intelligence Organisation attaché in Washington, Mervyn Jenkins, was caught passing AUSTEO (Australian Eyes Only) documents to the CIA. Jenkins was secretly monitored for months and eventually interviewed by DFA investigators on 11 June 1999. Two days later he committed suicide.[16]

Thanks to a *Four Corners* investigation we know that the Jenkins intelligence related to Indonesian involvement in escalating violence in East Timor. We also know that there was a policy dispute between Australia and the United States about the need for peacekeepers.

Did Australia withhold intelligence about the escalating violence in East Timor from its closest ally simply because it vindicated Roth's assessment? Surely there is a more substantial, national interest reason for Australia to risk antagonising its closest ally?

Australia's wilful blindness to the involvement of Indonesia's military in atrocities in East Timor continued up to the independence referendum.

On 30 August 1999, the people of East Timor voted 78 per cent in favour of independence.

On Sunday 3 September, when the result was officially announced, Indonesia launched a devastating campaign of destruction. Roth's assessment proved correct. Up to 2000 people were killed in the bloodbath that followed as Indonesian-backed militia killed independence sympathisers, burned homes and destroyed schools, hospitals and bridges.

CRITICAL MEETINGS

Following the referendum, Indonesia immediately lost all sovereignty over East Timor. The Timor Gap Treaty ceased to exist, again putting at risk the permits unilaterally issued by Australia north of the median line in the early 1960s.

A Senate Committee inquiring into Australia's relationship with East Timor and the Timor Gap Treaty held hearings in Darwin on 8 and 9 September 1999.[17] While East Timor was burning across the Timor Sea, James Godlove from Phillips (now ConocoPhillips) outlined the company's position in view of the 'material change' in East Timor's political status. Phillips was the operator of the Bayu-Undan gas field and was ready to invest $US1.4 billion in the first stage of a Darwin LNG plant. Godlove insisted the company could not wait for the new nation to formally come into existence. There was a Japanese buyer ready and the project needed certainty now. Under the Timor Gap Treaty, revenue from the field would have been split 50/50 between Indonesia and Australia. Phillips dangled the carrot of a desperately needed revenue stream for the new government.

On 20 September a United Nations peacekeeping force led by Australia left for East Timor to quell the devastating post-ballot violence. The day before, Gusmão had been released from house arrest in Jakarta. Australian officials advised it was unsafe for him to return to Timor.

Reluctantly Gusmão travelled to New York, where he and José Ramos-Horta met with Indonesian foreign minister Ali Alatas, who claimed events in East Timor had been beyond the control of the government. The Timor gap oil and gas fields were also discussed, and Alatas indicated Indonesia would hand over the Indonesian share of the Timor gap to the Timorese.[18] Gusmão arrived in Darwin on 6 October 1999, anxious to return to Dili, but a critical meeting had to take place before the Australian military deemed it safe for him to return.

The Timorese leadership had descended on Darwin, and Gusmão went from meeting to meeting with experts from around the globe, working on how to manage the enormous responsibility of administering a new country.

On 20 October 1999 Gusmão, accompanied by lawyer Mari Alkatiri and Ramos-Horta met with James Godlove from Phillips.[19] Godlove pitched the Bayu-Undan pipeline and Darwin LNG plant. He talked of billions of dollars that would flow to East Timor.

The traumatised Timorese leaders had no independent expert legal advice on the law of the sea. They were desperate to return to East Timor. They signed a statement saying they intended to 'negotiate appropriate transition arrangements and consequent changes in the current Treaty that maintain its legal authority over petroleum resource development'.[20]

The next day it was deemed safe for Gusmão to return. After dark on 21 October 1999, wearing his Falintil army uniform and accompanied by four Timorese bodyguards and Australian soldiers in full combat gear, he boarded a Royal Australian Air Force flight home over the Timor Sea.[21]

On 26 October 1999 a consortium led by Phillips announced that it would proceed with the $US1.4 billion investment in the first stage of the development of the Bayu-Undan field in the cooperation zone.[22] In a media release, Phillips said it had received a letter signed by Gusmão, Ramos-Horta and Alkatiri saying they would honour Timor gap petroleum zone arrangements.

*

Gusmão returned to a shattered nation. The people of East Timor were physically and psychologically destroyed. Families were starving, the medical and school systems non-existent, the power, water and road infrastructure in ruins.

In the period between the ballot in August 1999 and independence in May 2002, the United Nations Transitional Administration in East Timor (UNTAET) worked with the Timorese to establish the governance framework necessary for a new country. They faced a massive challenge. An electoral system had to be devised, a parliament and bureaucracy established. Decisions had to be made about what currency should be used, and what the official language should be. And the ownership of the Timor Sea oil and gas fields had to be sorted out with Australia.

Australia's agenda now was to get UNTAET, and newly independent East Timor, to step into the shoes of Indonesia and accept the terms of the Timor Gap Treaty. This was ambitious, given the UN and the Timorese leadership considered the Timor Gap Treaty illegal, and under the UN Convention on the Law of the Sea, the new nation of East Timor would have a right to an EEZ which would lead to a median line boundary.

Preliminary meetings between Australian and UN officials to discuss transitional arrangements in the Timor Sea were held in New York and Canberra during October and November 1999. There was debate within the UN about whether UNTAET had the mandate to negotiate a maritime boundary with Australia.[23] It was unprecedented for a UN transitional administration to negotiate with another state. The 1989 Timor Sea Treaty was dead, but Australia's diplomats were persuasive, and an interim arrangement was entered into on 10 February 2000, marked by an Exchange of Notes between Australia and UNTAET. The 1989 treaty's terms would continue, pending the negotiation of a new treaty between Australia and UNTAET prior to Timor-Leste's independence.

This was a remarkable diplomatic achievement for Australia, given that under international law in 2000 Australia had no rights in the Timor Sea north of the median line.

A NEW TIMOR GAP TREATY

While a traumatised people were drafting a constitution, and deciding a new national language, a parliamentary and electoral system, and attempting to establish the bare bones of a civil service, they were also having to negotiate a 'resource-sharing' treaty with multi-billion-dollar implications with one of the richest nations in the world. Australia pushed UNTAET to endorse an agreement closely resembling the terms of the Timor Gap treaty.

UNTAET's negotiators pushed back, insisting on a treaty reflecting current international law, i.e. the median line and wider lateral boundaries, including all of Greater Sunrise in the east and Laminaria and other fields in the west.

With Kafkaesque absurdity, Australia lodged a formal complaint with the UN secretary-general, arguing that the UN ought not be 'partisan' against one of its member states.[24]

Paul Cleary, a World Bank–appointed adviser to the first Timor-Leste government, detailed the relentless bullying tactics employed by Australia in the aftermath of the ballot in his book *Shakedown*. The meeting at which a deal was eventually struck in Canberra in mid-winter 2001 was a David and Goliath affair. The lead negotiator for UNTAET, former US diplomat Peter Galbraith, sat with Alkatiri, Ramos-Horta and four advisers on one side of the table. Opposite them sat Downer, Treasurer Peter Costello, Industry Minister Nick Minchin and forty grey-suited advisers.[25]

The result was a massive win for Australia.

The boundaries of the Zone of Cooperation Australia had negotiated with Indonesia were retained. This meant the Timorese only got a share of 18 per cent of revenue from Greater Sunrise field – 82 per cent would go to Australia. Similarly, Australia would continue to get 100 per cent of revenue from Woodside's Laminaria field, and other fields that lay just outside the western boundary. In 2001 Laminaria provided $81 million to the Australian economy.

Australia's one concession was to shift from a 50/50 to a 90/10 split (in favour of the Timorese) in the zone on the Timorese side of the median line. Australia was still claiming 10 per cent of revenue that under international law belonged to East Timor.

The deal had an even bigger sweetener for Australia. In addition to the 10 per cent of the revenue from Phillips' Bayu-Undan field (a field 100 per cent within East Timor's EEZ) the deal ensured Phillips' LNG plant in Darwin, estimated to be worth $25 billion to Australia over the next two decades, would go ahead.

The *Memorandum of Understanding of Timor Sea Arrangement* was signed at a ceremony in Dili on 5 July 2001 by Downer and Minchin, and Alkatiri on behalf of UNTAET. The agreement did not delimit a maritime boundary between Australia and Timor-Leste. Like the 1989 Timor Gap Treaty between Australia and Indonesia, the 2001 agreement was solely about how to carve up the oil and gas resources within the agreed boundary of a shared resource zone – so the Timor gap remained.

The maritime boundary between Australia and the new nation of East Timor would be negotiated according to the Law of the Sea Convention when East Timor formally existed.

*

Oceanic was still on the scene, and in early 2002 offered to fund an application by East Timor to the International Court of Justice for a ruling on a maritime boundary with Australia.

Australia was quick to respond. On 21 March 2002, Australia unilaterally withdrew from any international maritime boundary arbitration or judicial jurisdiction of the ICJ. Downer stated that, 'Australia's strong view is that any maritime boundary dispute is best settled by negotiation rather than litigation.' This left the Timorese unable to take Australia before an independent umpire to challenge Australia's claims beyond the median line to Sunrise, Troubadour, Laminaria and the other western fields. Alkatiri understatedly called this an 'unfriendly act'.

The Democratic Republic of Timor-Leste came into existence on 20 May 2002. As a result of the agreement reached between the UN and Australia, the first official act of the newly independent nation was to sign the heads of agreement for the 2002 Timor Sea Treaty with Australia.

Alexander Downer and Nick Minchin were there once again, their pens drawn for the signing ceremony.

AUSTRALIA VS. TIMOR-LESTE

Since Timor-Leste's independence Australia has continued to aggressively assert rights to oil and gas reserves that under international law would belong to Timor-Leste, and it has taken advantage of Timor-Leste's much weaker economic and political position to secure arrangements that benefit Australia to the disadvantage of the Timorese.

Timor-Leste declared its 200-nautical-mile EEZ in July 2002, a claim that overlapped Australia's permit areas north of the median line in the Timor Sea. Xanana Gusmão, who was now Timor-Leste's first president, told the Australian media, 'We are not asking for less or more than international law allows us to claim.' Timor-Leste's EEZ claim overlapped Greater Sunrise and Laminaria and nearby fields Buffalo and Corrallina.

In October 2002, Prime Minister Alkatiri wrote to Prime Minister John Howard proposing the commencement of maritime boundary negotiations as required under the Law of the Sea Convention. Australia insisted that the boundary negotiations would have to wait until all the agreements for the interim resource-sharing arrangements were in place. As discussed earlier, 18 per cent of Greater Sunrise was in what was now known as the Joint Petroleum Development Area (JPDA), and the rest in contested waters just outside the JPDA. It is not unusual in the petroleum industry for an oil or gas field to extend beneath national borders, in which case an agreement – known as a unitisation agreement – is negotiated, setting out a standard set of fiscal and regulatory arrangements to operate in the two jurisdictions. Australia wanted to negotiate the Greater Sunrise unitisation agreement before starting maritime boundary negotiations.

Alkatiri responded by publicly challenging Australia's right to what was then one of Australia's biggest oilfields,

Woodside's Laminaria, which lay just outside the western boundary of the JPDA in contested waters. Australia had already drawn US$1.2 billion ($1.7 billion) in royalties from Laminaria. Alkatiri said Timor-Leste would seek to recover past royalties paid to Australia.[26]

Downer flew to Dili on 27 November 2002 to get Timor-Leste to sign the Greater Sunrise unitisation agreement. He was accompanied by Ian Macfarlane, minister for industry, tourism and resources, his senior political adviser, Josh Frydenberg, and officials from the DFA.[27] The Timor Sea Office produced a transcript of the meeting that was later leaked to media site *Crikey*.[28]

Downer opened the discussion by accusing Alkatiri of attempting to renege on the 2002 Timor Sea Treaty:

> If I was in your position I would focus on revenue for your new and poor country and how to [progress] without compromising your integrity. To call us a big bully is a grotesque simplification of Australia. We had a cosy economic agreement with Indonesia, we bailed East Timor out with no economic benefit. Our relationship is crucially important, particularly for you, East Timor.

According to Paul Cleary, Downer pounded the table as he bluntly warned that Australia could leave all of the Timor Sea resources in the ground until it got its way. 'We don't have to exploit the resources. They can stay there for twenty, forty, fifty years.'[29] The transcript includes the following exchange:

> DOWNER: We are good and decent people and with very good faith. You say we ... shouldn't conclude an IUA [international unitisation agreement] before renegotiating boundaries. We can throw at you just as many lawyers to justify our boundary claim. Public opinion in Australia thinks 90 per cent is very generous.
> ALKATIRI: It is not with generosity that you gave us 90 per cent. We have lost 10 per cent.
> DOWNER: We claimed 100 per cent and we lost 90 per cent – I think that's a pretty good outcome for you.
> ALKATIRI: Don't get upset, please speak calmly on this issue. Our 100 per cent claim is based on international law and the equidistance line. It was not a random decision. The present issue of generosity – I do not accept.
> DOWNER: We had these negotiations two years ago.
> ALKATIRI: It is not generosity.
> DOWNER: We negotiated the 80/20 split a couple of years ago, if you are telling me you want to renegotiate —
> ALKATIRI: That is not what I am saying, why would we have initiated the ratification process? We are fulfilling our goals ... I am very pragmatic. The IUA is an economic issue but this doesn't mean we want to

> delay the discussions on maritime boundaries – this should be clear. Maritime boundaries should be drawn.
>
> ...
>
> DOWNER: You can't make us agree to your proposal ... You must understand, we won't agree to a new design on top of the existing TST [Timor Sea Treaty]. The joint development area on the Greater Sunrise is a change. That Morrow [a legal adviser for Timor-Leste] is very aggressive. Well, he has met his match with me. We won't agree to a JPDA for Greater Sunrise. We will do you due respect to listen to the proposal. We don't like brinksmanship. I think your Western advisers give you very poor advice that public opinion supports East Timor in Australia. We are very tough. We will not care if you give information to the media. Let me give you a tutorial in politics – not a chance.

Australia refused to ratify the Timor Sea Treaty before the Sunrise unitisation agreement was settled. This amounted to Australia threatening to delay approvals for Phillips' Bayu-Undan project, and therefore the date at which Timor-Leste would start receiving royalties desperately needed to fund health, education and infrastructure programs.

Timor-Leste eventually capitulated, after securing two concessions: Australia agreed to continue negotiations on a permanent boundary, and to put a pipeline from Greater Sunrise to Timor-Leste on the agenda. On 6 March 2002 Downer was back in the steamy heat of Dili in the Council of Ministers' room at the Palácio do Governo to sign the Sunrise unitisation agreement. Alkatiri refused to participate in the signing ceremony.

SPYING ALLEGATIONS

Talks to establish a permanent boundary began in Darwin in November 2003. Timor-Leste pressed for regular monthly meetings to resolve the dispute. Australia claimed it only had resources to meet twice a year.

Timor-Leste's foreign minister, José Ramos-Horta, told the National Press Club in Canberra in December 2003 that the delays in negotiating a maritime boundary were allowing Australia to exploit, under current licences, the Buffalo, Laminaria and Corallina oilfields, which rightfully belonged to Timor-Leste under international law.[30] He estimated US$1.5 billion ($2.03 billion) had gone to Australia under licences over the disputed fields since the mid-1990s.

Another round of talks was held in Dili in April 2004. The talks were extensively covered by Australian and international media. Galbraith led the delegation for Timor-Leste and Bill Campbell QC, from the attorney-general's department, led for Australia. Campbell made it clear that the disputed areas in Timor-Leste's 200-mile

EEZ 'were not up for negotiation'. Galbraith accused Australia of negotiating in bad faith: of attending a formal negotiation without having the disposition to negotiate.

Timor-Leste's legal adviser, Portuguese lawyer Nuno Antunes, pointed to more than sixty cases around the world where states less than 400 nautical miles apart had been settled by a simple median line. Australia insisted that Timor-Leste did not have a claim outside the JPDA. Three days of talks ended in stalemate. As Paul Cleary recalls, Alkatiri then did a 'marathon series of interviews that culminated in favourable coverage' in Australian and international media.

President Gusmão was also getting international coverage with comments comparing Timor-Leste's battle with Australia to the 'Timorese fight to free themselves from Indonesian domination'. Downer was furious. Cleary describes how he sent a secret envoy to Dili from the DFA in May to give Alkatiri and other senior government members the message that the Timor Sea issue could not be isolated from the bilateral relationship and the public commentary by the Timorese must stop. Downer had already decided to cut Timor-Leste's aid budget by 10 per cent in the 2004 Budget.

When Galbraith had accused the Australian delegation of bad faith, he was of course, unaware of allegations that surfaced in 2013 that the Australian government spied on the Timorese negotiating team. The allegations were made by a former ASIS agent (known as Witness K), who said that the Australian government authorised spies posing as aid workers helping with renovation work at the Palàcio do Governo in Dili to install electronic surveillance devices in the room being used by the Timor-Leste negotiating team. Downer had ministerial responsibility for ASIS at the time. The lawyer acting for Witness K emphasised that the alleged 'eavesdropping was no collateral product of a legitimate clandestine mission. The sole targeting was to capture clear voice transmission of out-of-session deliberations.'[31]

In an article in the *Australian* following the revelation of the spying allegations in 2013, Paul Cleary recalled that senior members of Timor-Leste's negotiating team 'believed one of their members was turned by ASIS during the 2004–05 negotiations over Greater Sunrise'.[32] As Gusmão explained, if the room was being bugged by Australia, the position of Timor-Leste's negotiators on key questions, their weaknesses on crucial points, ignorance of international legal issues and their bottom-line negotiating positions would have all been known in advance by Australia.[33]

In the lead-up to the October 2004 Australian federal election, supporters of Timor-Leste in Australia ran a very effective lobbying campaign putting pressure on the Howard government to recognise the median line as the permanent boundary and to give Timor-Leste a much bigger share in the whole of Greater Sunrise – i.e. in the

80 per cent outside the JPDA. Cleary says a breakthrough came at a meeting in August 2004, when Downer accepted Timor-Leste's claim outside the JPDA and made a cash offer rather than move toward a 50/50 share of Greater Sunrise. While the offer was rejected, it was enough to stop the Timorese leadership giving electorally unhelpful stories to the Australian media.

The Howard government was returned to power, but the Timor Sea issue didn't go away. Australia made another cash offer in January 2005. Again it was rejected. At a meeting in Dili at the end of April 2005, the Australian government 'offered' to give Timor-Leste a 50/50 share in the whole Greater Sunrise field. This offer was accepted.

The end result was the Treaty on Certain Maritime Arrangements in the Timor Sea, (CMATS), a 'temporary' revenue-sharing arrangement in the Timor Sea that included a fifty-year moratorium on maritime boundary talks. So the Timor gap remained – there was still no permanent maritime boundary between Australia and Timor-Leste.

CMATS was sold to the Australian public as a massive concession on Australia's part. Yet Australia's only concession was to increase Timor-Leste's share of Greater Sunrise to 50/50 – hardly generous when Greater Sunrise is 100 per cent on Timor-Leste's side of the median line. Australia's refusal to accept any movement on the boundaries of the JPDA, combined with the fifty-year moratorium on maritime boundary negotiations, provided most lucrative win. This outcome meant the contested Buffalo, Laminaria and Corallina fields on the western boundary would be fully exploited by Australia within the fifty-year timeframe, pouring billions of dollars into Australian coffers at Timor-Leste's expense.

The CMATS treaty was signed at a ceremony in Sydney in January 2007 by foreign ministers Alexander Downer and José Ramos-Horta. Soon after, DFA secretary Ashton Calvert retired. He joined the Woodside board eight months later. Downer retired from politics in 2008. He founded a boutique lobbying company, Bespoke Approach, and Woodside Petroleum became one of Bespoke's clients.

In Timor-Leste the challenge of rebuilding had almost became too much when rival sections of the army and the police came into conflict in 2006. The crisis brought down the Alkatiri government. Gusmão established a new political party and formed a coalition government. He was sworn in as prime minister in August 2007.

Gusmão didn't want to go the way of many resource-rich developing countries, where resources are exported, money pours in, an elite few get rich and the rest of the country goes backward. The Timor-Leste government started planning for a petroleum supply base on the south coast and argued the gas from Greater Sunrise should be piped there for processing.

Under CMATS, the jointly managed Timor Sea Designated Authority was scheduled to be replaced by a new Timor-Leste agency. Despite lobbying by Australia to delay the transition, the National Petroleum Authority was established on 1 July 2008. This meant the Woodside consortium now had to get approval from the Timor-Leste regulator, the National Petroleum Authority, and submit new documentation for approval.

Woodside refused to comply with a request from the National Petroleum Authority to lodge documentation not just on a floating plant, but for onshore-based LNG plants in Timor-Leste and Australia so the regulator could make a comparative assessment. This dispute was still playing out in 2012 when Gusmão learned of the CMATS spying allegations.

Gusmão immediately wrote to Australian prime minister Julia Gillard offering to confidentially deal with the spying allegations. The minister for resources and energy in Gillard's ALP government was Gary Gray, who had been Woodside's corporate affairs manager throughout the period CMATS was negotiated. The Australian government failed to respond to Gusmão's offer, so in April 2013 Timor-Leste initiated an arbitration dispute with Australia, seeking to have CMATS declared null and void on the basis of alleged espionage.

Gillard's government was defeated in June 2013 and the Liberal Party was back in power, led by Tony Abbott. On the eve of a preliminary hearing on the spying arbitration in The Hague in December 2013, ASIO and the Australian Federal Police simultaneously raided the office of the lawyer advising Timor-Leste, and the home of Witness K in Canberra. The raid, authorised by Attorney-General George Brandis, made international headlines. Timor-Leste immediately challenged Australia's actions in the ICJ. In March 2014 the court ordered Australia to seal the seized documents and keep them sealed until its final decision and to desist from interfering 'in any way in communications between Timor-Leste and its legal advisers in connection with the pending Arbitration'.

In July 2015 the ALP National Conference supported a motion moved by long-time Timor-Leste campaigner Janelle Saffin to begin maritime boundary negotiations with Timor-Leste if Labor were to regain power. There was no debate. Steve Bracks sat in the back stalls and applauded the change in policy.

TIMOR-LESTE TAKES AUSTRALIA TO UN COMPULSORY CONCILIATION

By then Gusmão had won another election in 2012 and controversially handed power to Opposition member Dr Rui Maria de Araújo. Gusmão stayed in the government as the minister for planning and strategic investment.

In February 2016 Prime Minister Araújo wrote to the

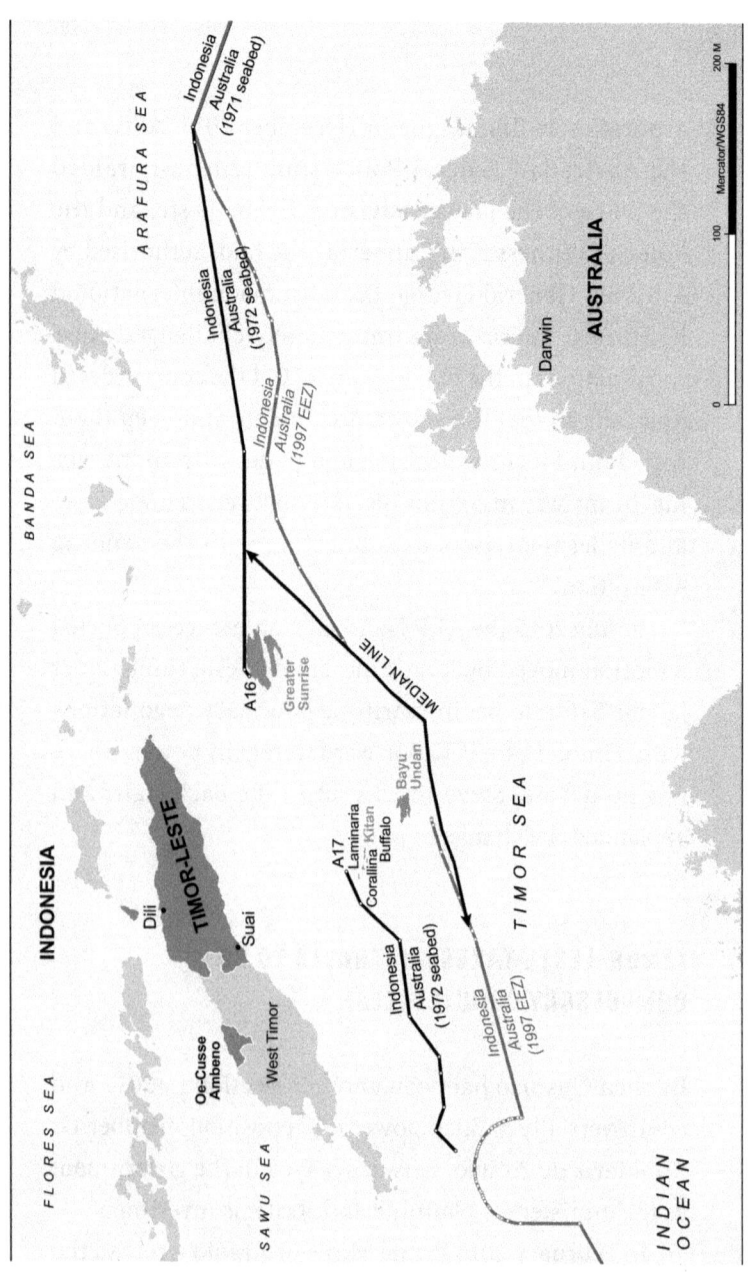

Figure 3

new prime minister of Australia, Malcolm Turnbull, seeking to commence negotiations on permanent maritime boundaries. When Australia rejected the offer, Timor-Leste exercised its right to institute compulsory conciliation under the UN Convention on the Law of the Sea. This process, which is being used for the first time, was specifically conceived for situations where there is a dispute regarding maritime borders and one party has withdrawn from binding dispute resolution procedures under the Convention.

At the federal election on 2 July 2016 Turnbull hung on to power by one seat. This gave Timor-Leste no choice but to press on with the Compulsory Conciliation.

The first hearing on 29 August 2016 was public. Xanana Gusmão presented a historical summary of Timor-Leste's position and Timor-Leste's lawyers, law of the sea experts Vaughan Lowe QC and Sir Michael Wood, outlined the claim to the median line.

Australia's submission continued the DFA mythmaking. Its lawyers asserted that Australia was 'instrumental in securing international support for the referendum process that led to independence'. The submission then claimed Australia sought to negotiate maritime boundaries in the CMATS negotiations, and it was Timor-Leste that expressed a preference to negotiate a resource-sharing agreement. What they failed to say was that this only happened after Australia made it clear that the only boundary line it was prepared to negotiate was a straight line to close the gap, and that it would not accept any boundary proposal outside the JPDA.[34]

Extraordinarily, the submission also repeated the DFA myth that the Timor Trough means the 'physical continental shelves of Australia to the south and Timor-Leste and Indonesia to the north are entirely separate'.[35] The government had geological advice even before the negotiations started with Indonesia that this was not the case, and geological analysis since has concluded that Australia's continental shelf extends beneath the island of Timor.

Australia immediately challenged the jurisdiction of the commission. That debate was held behind closed doors. On 26 September 2016, the Conciliation Commission found against Australia on the jurisdiction issue on all counts.

A second round of hearings held in January 2017 led to a significant breakthrough – Australia agreed to work in good faith towards an agreement on maritime boundaries by the end of the conciliation process in September 2017. In return Timor-Leste agreed to drop the embarrassing spying arbitration and a separate taxation-related arbitration.

While Australia's participation in the process is compulsory, the outcome of the process is not binding on either party.

There is a glimmer of hope, however, that for the first time in over half a century, the diplomats in the DFA and their political masters may have finally worked out that a permanent median line boundary in the Timor Sea, based on international law, is in Australia's national interest.

CONCLUSION

Two documents in the DFA files stayed with me. Both were written by Richard Woolcott, and involved him making the case that pragmatism, rather than idealism, was in Australia's national interest.

In the first cable, from August 1975, he argued that closing the Timor gap could be 'much more readily negotiated with Indonesia' ... 'than with Portugal or independent Portuguese Timor', acknowledging that, 'I know I am recommending a pragmatic rather than a principled stand but this is what national interest and foreign policy is all about.'

The second was his 'Kissingerian realism' cable from early January 1976. In this long missive to the new Fraser government he confidently advised against taking a 'moral stance' and supporting the right of the people of East Timor to self-determination because a 'pragmatic and realistic acceptance of the longer term inevitabilities of the situation' was in Australia's national interest.

The first cable was memorable because it was one of the few that articulated Australia's Timor Sea oil agenda in the declassified files. It would probably still be classified if it hadn't been leaked in 1980.

The second was memorable for its naked abrogation of morality, for the influence it had on Australian policy, and for the number of assumptions that have since been proved wrong.

'National interest' is a commonly used – and abused – phrase for diplomats and politicians.

Garfield Barwick, for example, believed it was in Australia's national interest to reject the United States' request for Australia to take a lead role in Kennedy's 'preventative' diplomacy in Portuguese Timor.

In 1965 David Fairbairn argued a median line boundary in the Timor Sea was in Australia's national interest as it would lower the risk of conflict with Indonesia and Portugal and provide security for the permit holders.

Robin Ashwin argued it was in Australia's national interest to have a good long-term relationship with Indonesia and Portuguese Timor, and that required 'drawing a median line in the middle of the sea between us and them'.

Ambassador Furlonger doubted if McMahon's hard-line negotiating strategy with Indonesia in 1972 was in Australia's national interest.

Don Willesee, as foreign minister before the invasion, pushed for the Timorese to have a UN-sanctioned vote of self-determination, because it was in Australia's national interest to support the United Nations.

Senior figures in the Department of Defence believed an independent East Timor was in Australia's national interest as a friendly base in case of future hostilities with Indonesia.

And following the Indonesian invasion there were individuals at senior levels in the government so incensed with Woolcott's enthusiasm for 'Kissingerian realism' that they leaked that cable and many others to the *Canberra Times* in 1976.

More recently there was Witness K, who blew the whistle on Australia's alleged bugging of the Timorese during the CMATS negotiations. Witness K believed the 2004 bugging operation was '"immoral and wrong" because it served not the national interest, but the interests of big oil and gas'.[1]

The records show that Woolcott's interpretation of Australia's national interest prevailed. The sentiments he voiced and the strategies he set out were adopted by the Whitlam, Fraser, Hawke and Howard governments. For the next three decades and beyond, Australia eschewed a 'moral stance' as it pragmatically pursued its objective to claim oil and gas reserves north of the median line in the Timor Sea.

At no point did Australia's officials decide that enough was enough, that it was no longer in Australia's national interest to condone Indonesia's brutal invasion and occupation. Not when it became clear that Fretilin was putting up significant resistance, and that the military engagement was going to be prolonged. Not when the files started filling up with intelligence and other reports about the brutality of the invasion and the rising death toll that was due in part to a deliberate policy to 'starve out' the resistance. Not when legal advice confirmed that Australia would need to give de jure recognition to the occupation in order to negotiate to close the Timor gap.

Instead, in the Fraser years the DFA doubled down. The elite of Australia's bureaucracy, the most highly educated and brightest men of their generation (in the '60s and '70s they were invariably men), were so committed to Kissingerian realism that they covered up reports of massacres, torture and mass starvation in pursuit of what they perceived to be Australia's national interest.

The deaths and atrocities continued in East Timor, and the cover-ups continued under Hawke. Australia's immorality under Hawke climaxed in December 1989 when Gareth Evans and Ali Alatas, to quote Xanana Gusmão, 'signed an agreement to exploit our resources, drinking champagne and forgetting that we were fighting there'.[2]

There was a new generation of officials at the policy helm by the late '90s when, in one of the least recognised and most despicable acts of diplomatic bastardry in the DFA Timor Sea playbook, the Howard government sought to maintain rights to oil and gas north of the median line, irrespective of international law and the rights of the Timorese. Again, this policy was considered to be in Australia's national interest.

Australia succeeded in forcing the UN and a devastated and exhausted Timorese leadership to agree to continue the terms of the defunct Timor Gap Treaty, and then allegedly spied on the Timorese team during the CMATS negotiations. And when the spying allegations became public we had the bad-faith arbitration and the ASIS and AFP raids on Timor-Leste's lawyer and key witness, and Timor-Leste's success in the International Court of Justice. It is hard to see how any of this has been in Australia's national interest.

The Timor gap has still not been closed.

There is only a permanent seabed boundary between Australia and Timor-Leste.

There is no permanent maritime boundary between Australia and Indonesia in the Timor Sea and the Arafura Sea. Indonesia has refused to ratify a water column boundary because its parliament does not want to be reminded just how much Indonesia conceded in 1972. It is ironic that the area of Australia's massive coastline of most concern to the Australian Border Force is the one with no ratified international maritime border.

Some forty-three years after discovering Greater Sunrise, Woodside still does not have the security of tenure necessary to commence exploiting the field.

Timor-Leste is today diplomatically closer to Indonesia and China than it is to Australia.

Australia is unable to lend credible support to the United States and its ASEAN neighbours in their dispute with China in the South China Sea because Australia's 'continental shelf' claim in the Timor Sea is on a par with China's nine-dash-line claim beyond the median line in the South China Sea.

Since the spying allegations in 2013, the relationship between Australia and Timor-Leste is so fraught that no federal government minister has visited Timor-Leste.

However, in strictly economic terms, the DFA's Timor Sea agenda has delivered big time. Australia has collected billions of dollars in revenue from permits issued north of the median line in waters that, according to international law, fall within Timor-Leste's EEZ. In addition, the ConocoPhillips LNG plant in Darwin has contributed more than $25 billion to the Australian economy.

*

The DFA has gone to extraordinary lengths to play down Australia's oil agenda in the Timor Sea and its influence on Australia's foreign policy. But while the DFA might have succeeded in writing Australia's interest in oil in the Timor Sea out of Australia's history books, in Timor-Leste there is a high level of community awareness of the issue. The streets of Dili are graffitied with kangaroos leaping away with buckets of oil.

Australia's ruthless pursuit of its secret Timor Sea oil agenda over the last fifty years leaves little room for optimism. Over and over again, Australia has crossed the line.

However, for the first time Australia is having to defend its actions before an independent panel. The UN's Compulsory Conciliation hearing may prove to be a game changer.

Xanana Gusmão is a fighter, the Timorese people are resilient and determined, and international law is on Timor-Leste's side. Perhaps we should recall the words of Martin Luther King Jr: 'The arc of the moral universe is long, but it bends towards justice.'

A NOTE ON THE NATIONAL ARCHIVES OF AUSTRALIA

The National Archives of Australia (NAA) was established in 1983 to 'preserve Australia's most valuable government records and encourage their use by the public; and to promote good records management by Australian Government agencies'. The *Archives Act 1983* gives all Australians a legal right of access to Australian government records when they enter the 'open access period', subject to certain exemptions. Since 2010, all documents created in the 1970s have been in the open access period.

Many of the files relevant to Australia's Timor Sea oil agenda had not previously been accessed, which meant they had to be examined by archives staff and the responsible department, usually the DFA, for any information that 'could adversely affect Australia's defence, security or international relations'. If the NAA has not made a decision in ninety days one has an automatic right of appeal to the Administrative Appeals Tribunal (AAT). I had dozens of requests over the ninety-day mark, so in October

2014 I paid $800 and sought a formal review of the NAA's failure to make a decision in regard to twenty-three files concerning Australia's Timor Sea boundary negotiations between 1972 and 1979.

The Australian government's ongoing sensitivity about its Timor Sea oil agenda has made dealing with the NAA a very interesting process, involving hundreds of pages of files with tantalising redactions, a mysterious 'lost' file and 'inadvertently' released material.

THE LOST FILE

In April 2015 the NAA advised that one of the files in the batch I had appealed to the AAT 'cannot be located'. The file in question had been requested by another researcher in 2009 and assessed by the DFA in 2011. It was not returned to the NAA and, following my request, could not be found at the DFA or the archives.

The NAA concluded that 'in these unfortunate and unusual circumstances' the file was 'not in the care of the Archives', nor in 'the custody of DFAT', which meant that I should withdraw the file from the AAT appeal.[1] Based on this reasoning the entire archives collection could be sent off to the DFA and never come back – with no consequences.

I sought the advice of the AAT. At a hearing on 23 June 2015 the Australian Government Solicitor lawyer and NAA legal adviser told the tribunal that the file had been classified 'secret', which meant it would have been subject to special handling procedures. All staff in relevant areas had been asked to search for it. Attempts made to contact ex-DFA staff to see if they had any recollection of the file 'drew a blank'. I declined to withdraw the file from my appeal and the AAT agreed the issue could be addressed when the matter finally came to hearing.

Then miraculously, in mid-February 2016, the file was located in a safe. According to the NAA, the safe was searched as part of the DFA's 'extensive searches for the file' but was 'not located on those occasions. Recently the safe was being emptied of its entire contents, and the file was found in a little-used drawer, wedged underneath unrelated files and out of sight.'

The file was assessed and eventually made available for me to inspect in September 2016. There were seventeen pages with redactions and the documents between pages 183 and 210 removed entirely. Among other fascinating material, the file included the negotiating brief for the first round of maritime boundary talks with Indonesia in February 1979. Even with the redactions, it was a treasure trove.

THE INADVERTENTLY RELEASED FILE

During my visit to the NAA in September 2016 I reviewed one of the files I was appealing, as the NAA had declassified some additional material. My modus operandi was to laboriously photograph files in chronological order on my phone – 100, 200, sometimes 300 pages – so I could review them properly later.

This file covered the 1979 negotiations and was still heavily redacted. The night after I reviewed this file in the reading room, I received a message to call the Australian Government Solicitor urgently. I called in the morning and, clearly very flustered, the solicitor explained that I might have been given access to a page with classified material that had been 'inadvertently' released. I said I'd need some time to get legal advice.

Then the registrar at the AAT called to tell me I had to participate in an urgent directions hearing that afternoon. It was Kafkaesque. I reserved my right to revisit the issue when the matter finally came to hearing and in the meantime was required to destroy any copies I had made of the offending page and 'prohibited from disclosing the contents ... in the form in which it was made available'. I was later emailed the offending page, with the inadvertently released half-sentence appropriately redacted.[2]

The page is from the brief for the February 1979 negotiations with Indonesia.

I cannot un-see what I read on that file. While I cannot disclose the part sentence redacted in paragraph 26, I can say that the redaction suggests a sensitivity about references to a certain phrase bordering on paranoia.

16.

The Western Sector

26. As noted in paragraph 25 above, the western sector of the Australian continental shelf (the Scott Plateau) is separated from the Indonesian shelf by the Java Trench - Timor Trough geological boundary. This runs to the north of where a median line continuation of the present agreed boundary would be located. Although Cabinet has directed that Australia could move [REDACTED s 33(1)(a)] it would be consistent with our approach to the gap and 1972 negotiations if we took as our starting point the edge of the Australian continental shelf in the western sector and negotiated back from there.

Preferred Australian solution

[REDACTED s 33(1)(a)] Closure of the gap by a straight line joining the ends of the present boundary would be our preference, because:

(a) it would give us the larger part of what is a prospective area;

(b) it would involve only minor disruption of the present Australian exploration permits;

(c) it would facilitate the negotiations.

.../17

ACKNOWLEDGMENTS

I had the opportunity to work with Steve Bracks AC in Timor-Leste thanks to the generous support of Harold Mitchell AC. Steve and Harold share a passion for Timor-Leste that inspired my interest and I cannot thank them enough.

Thanks to Rosa Silvestro, who makes everything happen, and the supervisors of my Ph.D. in progress at Monash University, Associate Professor Paul Strangio and Dr Sara Niner, whose remarkable biography of Xanana Gusmão haunted me as I read the archives. Thanks to Professor Clinton Fernandes, who helped me navigate the National Archives and provided invaluable advice, and to James Dunn, Kirsty Sword Gusmão, Jill Jolliffe, Robert King and John Waddington for paving the way. Joel Deane, Petrina Dorrington, Shelley Penn and Dr Carolyn Woodley read early drafts and pushed me to keep going.

Thank you to the team at Black Inc., especially Morry Schwartz for his enthusiasm, Chris Feik, and my editor, Jo Rosenberg, who transformed my research into readable prose. Jo coaxed this book out of me and made the impossible doable.

Thanks to the dedicated advisers, Timorese, Portuguese and Australian, I have had the pleasure of working with over the years. And to the extraordinary leaders of Timor-Leste: I am in awe of the commitment of Minister Alfredo Pires, the intellectual force of Minister Agio Pereira and the strategic genius of Xanana Gusmão.

And finally, thanks to my family, Bruce and Emma, who keep the joy alive.

ENDNOTES

Introduction

1. Paul Cleary, *Shakedown: Australia's Grab for Timor Oil*, Crows Nest: Allen & Unwin, 2007.
2. Wendy Way (ed.), *Australia and the Indonesian Incorporation of Portuguese Timor 1974–1976*, Documents on Australian Foreign Policy, Canberra: Melbourne University Press and DFAT, 2000, http://dfat.gov.au/about-us/publications/historical-documents/Documents/australia-and-the-indonesian-incorporation-of-portuguese-timor-1974-1976.pdf.
3. The Australian government department responsible for international relations was known as the Department of External Affairs until 6 November 1970, when its title was changed to the Department of Foreign Affairs. On 24 July 1987 its name was changed to the Department of Foreign Affairs and Trade. To avoid confusion, and consistent with the Downer Compilation, 'DFA' is used throughout this book.
4. Policy Planning Paper, 3 May 1974, Downer Compilation, p. 50. Original on NAA:1757393, DFA, *Regional Defence – Defence of Pacific – East and South East Asia Timor,* A1838 696/5 Part 2.
5. Clinton Fernandes, 'Accomplice to mass atrocities: The international community and Indonesia's invasion of East Timor', *Politics and Governance*, vol. 3, no. 4, 2015.
6. The paragraph was first published in Bruce Juddery, 'Envoy puts Jakarta's view', *Canberra Times*, 16 January 1976; see also 'Cablegram to Canberra, Jakarta, 5 January 1976', Downer Compilation, p. 657.

Chapter 1

1. Robert Murray, *From the Edge of a Timeless Land: A History of the North West Shelf Gas Project*, North Sydney: Allen & Unwin, 1991, p. 8.
2. Peter Purcell, Mike Butcher & Yolande M.J. Collins, 'Nicholas Boutakoff and Australia's North West Shelf', *Search and Discovery*, 2015, http://www.searchanddiscovery.com/documents/2015/70152purcell/ndx_purcell.pdf.
3. S.V. Scott, 'The inclusion of sedentary fisheries within the continental shelf doctrine', *The International and Comparative Law Quarterly*, vol. 41, no. 4, 1992.
4. *Harlowe's Nominees P/L v. Woodside (Lakes Entrance) Oil Co. Ltd. and Burmah Oil Australia Limited* [1967] VicSC 130, 25 August 1967.
5. Graeme Atherton, *Fifty Years of Woodside's Energy*, Woodside Petroleum Ltd, 2004, p. 13.
6. Atherton, *Fifty Years of Woodside's Energy*, p. 14.
7. Cabinet Submission No. 1165, 'Off-shore petroleum: Legislation to give effect to joint commonwealth–state legislative arrangements', 25 November 1965, Prime Minister's Department, *Off-Shore Minerals – Commonwealth/State discussions*, NAA barcode: 1345145, A4940, C3945. Barwick served as Minister for External Affairs from 1961 to 1964. He led the Australian Delegation to the 1960, 1962 and 1964 sessions of the General Assembly of the UN. He resigned in 1964 to become chief justice of the High Court of Australia.
8. Nicholas Boutakoff, 'Geology of the off-shore areas of North-Western Australia', *Australian Petroleum Exploration Association Journal*, 1963, p. 10.
9. Robert J. King, 'A gap in the relationship: The Timor Gap, 1972–2013', submission to the Inquiry into Australia's Relationship with Timor-Leste Joint Standing Committee on Foreign Affairs, Defence and Trade Foreign Affairs Subcommittee, Parliament of Australia, Canberra, 2013.
10. Barwick to Beale, 5 February 1963, DFA, *Other Non Self-Government Territories – Portuguese Timor*, NAA barcode: 546777, A1838 935/17/3 Part 1.
11. 'Discussion on Portuguese Timor', 12 February 1963, DFA, *Indonesia – Relations with Portuguese Timor Part 3*, NAA barcode: 547385, A1838 3006/4/3.

12	Cabinet Minutes, 5 February 1963, Decision No. 632, Indonesia – Quadripartite Talks, Prime Minister's Department, *Indonesia – Quadripartite Talks in Washington*, NAA barcode: 1338483, A4940, C3739; Nicholas Tarling, *Britain and Portuguese Timor 1941–1976*, Monash Asia Series, Clayton, Vic.: Monash University Publishing, 2013.
13	Cabinet Submission, 'Portuguese Timor', 21 February 1963, DFA, *Portuguese Timor*, NAA barcode: 546778, A1838 935/17/3 Part 2.
14	Barwick to Rylah, 25 September 1963, Attorney-General's Department, *Continental Shelf – Proposed Commonwealth Offshore Oil Legislation*, NAA barcode: 1110524, A432 1963/3189.
15	Cabinet Submission No. 1165, NAA: A4940 C3945.
16	Vincent Smith, 'Australia risks oil challenge by Indonesia', *Australian*, 20 December 1965.
17	Transcript, Robin Ashwin interview by Sara Dowse, 25–26 October 2006, National Library of Australia, p. 83.

Chapter 2

1	Laurie Oakes interview with Alex Malley, *Conversations with Alex Malley*, Channel 9, aired 7 February 2016.
2	Department of National Development, 'Law of the Sea, Notes for Discussion with U.S. delegation', 23 November 1971, Prime Minister's Department, *Law of the Sea – Application to Australia 1971*, NAA barcode: 3161700, A5882 CO567 Part 3.
3	DFA, *Indonesia–Australia Continental Shelf Boundary Negotiations*, NAA barcode: 558528, A1838 752/1/23 Part 1.
4	Among other papers, Sandiford and Duffy referred me to M.G. Audley-Charles, D.J. Carter & J.S. Milsom, 'Tectonic development of Eastern Indonesia in relation to Gondwanaland dispersal', *Nature Physical Science*, vol. 239, 18 September 1972; M.G. Audley-Charles, 'Tectonic post-collision processes in Timor', London: Geological Society, Special Publications, 2011.
5	'The Timor trough – A summary of current geological knowledge', Bureau of Minerals Resources, February 1970, DFA, *Australia–Portuguese Negotiations on the Continental Shelf Annex*, NAA barcode: 558645, A1838 756/1/4.

6 'Opening statement by Sir Laurence McIntyre, leader of the Australian delegation', 19 March 1970, DFA, *Jakarta – Political: Continental Shelf*, NAA barcode: 4209043, A4359 201/2/3 Part 2.
7 Justice Gleeson, Solicitor-General, Opening Session, Permanent Court of Arbitration Case No 2016-10, *Conciliation Proceedings between the Government of the Democratic Republic of Timor-Leste and the Government of the Commonwealth of Australia Pursuant to Article 298 and Annex V of the UN Convention on the Law of the Sea*, 29 August 2016, pp. 92–93. Online at https://pcacases.com/web/sendAttach/1889.
8 Lisbon to Secretary, DFA, 21 August 1970, Attorney-General's Department, *Portuguese Timor – Continental Shelf*, NAA barcode: 11089375, A5034 SR1974/3009 Part 1.
9 Portuguese Ministry of Foreign Affairs, Directorate-General of Economic Affairs, to Australian Embassy Lisbon, 2 November 1970, NAA barcode: 558641, DFA, *Australia Portugal Negotiations on Portuguese Timor Continental Shelf*, A1838 756/1/4 Part 3.
10 Brennan to Bury, Timor Oil Concessions, 7 April 1971, DFA, 'Australia Portugal Negotiations on Portuguese Timor – Continental Shelf', NAA barcode: 558637, A1838 756/1/4 Part 1.
11 Quoted in King, 'A gap in the relationship', p. 4.
12 Minute, Bureau of Mineral Resources, Geology and Geophysics, Department of National Development, 6 July 1971, NAA: A5034 SR1974/3009 Part 1.
13 Brennan to Minister for Foreign Affairs, 'Burmah Oil Company: Drilling in Timor Sea', 25 August 1971, DFA, *Indonesia – Australia – Indonesia – Continental Shelf Boundary Negotiations*, NAA barcode: 558539, A1838 752/1/23 Part 8.
14 Brennan to Minister for Foreign Affairs, 25 August 1971, NAA: A1838 752/1/23.
15 *Royal Commission on Intelligence and Security Fifth Report [Re. Australian Secret Intelligence Service] – Volume I (Copy No 25) – [Reference Copy]*, NAA barcode: 30091092, A8908 5A.
16 John G. Butcher, 'Becoming an archipelagic state: The Juanda Declaration of 1957 and the "struggle" to gain international recognition of the archipelagic principle', in Robert Cribb & Michele Ford (eds), *Indonesia Beyond the*

Water's Edge: Managing an Archipelagic State, Singapore: Institute of Southeast Asian Studies, 2009, p. 39. Indonesia shares a land border with Malaysia, Papua New Guinea and Portuguese Timor, and sea borders with ten countries, more sea borders than any other country in the world.

17 Donald Rothwell et. al. (eds), *The Oxford Handbook of the Law of the Sea*, Oxford: Oxford University Press, 2015, p. 140.

18 Rowan Callick, 'Tiny Timor treads warily among giants', *Australian Financial Review*, 31 May 2004. Such a side deal is also consistent with reports that Indonesia allowed Malaysia a larger seabed area than Indonesia in the Malacca Strait as a 'gift' for their support in 'pushing the regime of Archipelagic Waters during negotiations for the Third United Nations Law of the Sea Conference': Leonardo Bernard, 'Whose side is it on? The boundaries dispute in the North Malacca Strait', *Indonesian Journal of International Law*, vol. 9, no. 3, 2012, p. 388.

19 Record of conversation, Malik and Bowen, 7 February 1972, DFA, *Indonesia – Political Relations with Australia – General*, NAA barcode: 545462, A1838 3034/10/1 Part 34.

20 'Cabinet submission, Sea-Bed Boundary with Indonesia', May 1972, DFA, *Australia–Indonesia Continental Shelf Boundaries – (Agreement) Negotiations*, NAA barcode: 551873, A1838 752/1/23 Part 11. The first paragraph in this submission is redacted on national interest grounds.

21 'Cabinet Minute Decision 999 on Submissions 674 and 675', 23 May 1972, DFA, *Australia–Indonesia Continental Shelf Boundaries – (Agreement) Negotiations,* NAA barcode: 551878, A1838 752/1/23 Part 14.

22 Livermore to DFA, 8 September 1972, DFA, *Law of the Sea – Oil Mining on the Australian Continental Shelf*, NAA barcode: 1764975, A1838 1733/5 Part 1.

23 King, 'A gap in the relationship', pp. 6–9.

24 DFA Background Paper, 'Australia's relations with Indonesia', 12 February 1973, DFA, *Jakarta – Relations with Australia,* NAA barcode: 4227204, 821/1/1 Part 5.

25 King, 'A gap in the Relationship', p. 8; Michael Richardson, 'Tying up Timor's loose ends', *Far Eastern Economic Review*, 5 January 1979, p. 44; Hamish McDonald, 'Settling the maritime borders with Timor-Leste', *Saturday Paper*, 20 May 2017.

CHAPTER 3

1. Rodney Tiffen, *Diplomatic Deceits: Government, Media and East Timor*, Sydney: UNSW Press, 2001, p. 7.
2. Referred to in Backen to Noakes, 4 September 1975, Department of National Resources, *Australia Portuguese Timor: Seabed Boundary: Timor Gap Treaty*, NAA barcode: 31167593, A1690 DPIE1975/958.
3. Hewitt to Secretary Attorney-General's Department, 22 November 1973, NAA: A5034 SR1974/3009 Part 1.
4. Cable from Kelly to Canberra, 11 November 1973, NAA: A5034 SR1974/3009 Part 1.
5. Brief Loomes to Connor, 28 November 1973, *Australia Portugal Negotiatons on Portuguese Timor – Continental Shelf*, NAA barcode: 558639, A1838 756/1/4 Part 2.
6. Hewitt to Secretary Attorney-General's Department, 23 August 1974, NAA: A5034 SR1974/3009 Part 1.
7. Nuno Sergio Marques Antunes, *Towards the Conceptualisation of Maritime Delimitation: Legal and Technical Aspects of a Political Process*, Leiden, Boston: Martinus Nijhoff, 2003, p. 357.
8. Loomes to Acting Minister Foreign Affairs, 21 March 1974, DFA, *Australia Portugal Negotiations on Portuguese Timor Continental Shelf*, NAA barcode: 558641, A1838 756/1/4 Part 3.
9. Transcript, 'Prime Minister's interview on Perth television (on 25 March) relating to oil leases on the North West Shelf', NAA: A1838 756/1/4 Part 3.
10. *Australian*, 27 March 1974, transcript on NAA barcode: A1838 756/1/4 Part 3.
11. 'Whip-crack at Portuguese', *Age,* 28 March 1974.
12. Embassy of Portugal to DFA, 18 April 1974, NAA: A5034 SR1974/3009 Part 1.
13. Policy Planning Paper, 3 May 1974, Downer Compilation, p. 50.
14. Policy Planning Paper, 3 May 1974, DFA, *Regional Defence – Defence of Pacific –East and South East Asia Timor,* NAA barcode: 1757393, A1838 696/5 Part 2.
15. Smith to Landale, 'Timor oil exploration', 6 July 1974, DFA, *Australia–Portugal – Negotiations on Portuguese Timor – Continental Shelf,* NAA barcode: 558643, A1838 756/1/4 Part 4.

16 Peter Hastings, 'Voice for Timor independence', *Sydney Morning Herald*, 22 July 1974.

17 Hamish McDonald, 'It's tiny, poor, and very possibly not going to take it anymore', *Global Mail,* 28 March 2013.

18 'Woodside hit gas off Timor', *Age,* 28 August 1974.

19 Hewitt to Renouf, 23 August 1974, NAA: A5034 SR1974/3009 Part 1.

20 Brazil to Secretary Minerals and Energy, 13 September 1974, NAA: A5034 SR1974/3009 Part 1.

21 Richard Woolcott, *The Hot Seat: Reflections on Diplomacy from Stalin's Death to the Bali Bombings*, Pymble, NSW: HarperCollins, 2003, p. 151.

22 Jusuf Wanandi, *Shades of Grey: A Political Memoir of Modern Indonesia, 1965–1998*, Jakarta: Equinox, 2012, p. 195.

23 Wanandi, *Shades of Grey*, p. 196.

24 Hugh Armfield, 'Canberra aim for Timor: Go Indonesian', *Age*, 13 September 1974.

25 Gilchrist to Coles, 9 July 1974, NAA: A1838, 756/1/4 Part 4.

26 Body to Secretary Attorney-General's Department, 9 August 1974, NAA: A5034 SR1974/3009 Part 1.

27 DFA Canberra to DFA New York, 26 September 1974, NAA barcode: 11089375.

28 Douglas Wilkie, *Sun-News Pictorial*, 10 September 1974, article transcribed on *Portugal – Foreign policy – Portuguese Timor*, NAA: 591530, A1838 49/2/1/1 Part 3.

29 'Summary of meeting to consider outer limits of outer edge of Australia's continental margin', draft, 8 January 1974, NAA: A1838 1733/5 Part 1.

30 G.G. Barnden, 'Oil situation in East Timor', 7 December 1978, DFA, *Law of the Sea – Delimitation – Australia – Indonesia (Timor),* NAA barcode: 1872070, A1838 1733/3/2 Part 5.

31 James Cotton, 'Australia's commitment in East Timor: A review article', *Contemporary Southeast Asia*, vol. 23, no. 3, 2001, p. 553.

32 See Adam Hughes Henry, *The Gatekeepers of Australian Foreign Policy*, North Melbourne: Australian Scholarly Publishing, 2015, pp. 148–55.

33 Lisbon to Canberra, 8 April 1975, 'Oil Prospecting in the Timor Sea, Department of National Development,

Australia Portuguese Timor: Seabed Boundary: Timor Gap Treaty,' NAA barcode: 31167593, DPIE1975/958.

34 'Briefing by Dr Avelar Barbosa (mining and petroleum technical expert)', 26 June 1975, DFA, *Law of the Sea – Delimitation Australia Portugal Timor,* NAA barcode: 1763283, A1838 1733/3/3 Part 2.

35 Oceanic to Hewitt, 28 July 1975, NAA: A1690 DPIE1975/958.

36 Barnden, 'Oil situation in East Timor', NAA: A1838 1733/3/2 Part 5; See also Hamish McDonald, 'Indonesia cool on Timor oil search', *Australian Financial Review,* 29 December 1975.

37 Scully to Singer, 28 August 1975, NAA: A5034 SR1974/3009 Part 1.

38 Woolcott to Secretary DFA, 17 August 1975, Downer Compilation p. 314; Paras 3, 4, 10 and 11 were leaked to Laurie Oakes and published in the *Sun* on 1 May 1976.

39 Referred to in Hewitt to Brennan, 21 February 1975, NAA: A1690 DPIE1975/958.

40 Backen to Noakes, 4 September 1975, NAA: A1690 DPIE1975/958.

41 Noakes, to Backen, 13 October 1975, NAA: A5034 SR1974/3009 Part 1.

42 Desmond Ball, 'Silent witness: Australian intelligence and East Timor', *Pacific Review,* vol. 14, no. 1, 2001.

43 Malcom Fraser and Margaret Simmons, *Malcom Fraser: The Political Memoirs,* Melbourne: Miegunyah Press, 2010, p. 541.

44 Sara Niner, *Xanana: Leader of the Struggle for Independent Timor-Leste,* North Melbourne: Australian Scholarly Publishing, 2009, p. 29.

45 Niner, *Xanana,* p.30.

46 Wanandi, *Shades of Grey,* p. 211.

47 James Dunn, *East Timor: A Rough Passage to Independence,* Double Bay, NSW: Longueville Books, 2003, p. 243.

48 Sarah Niner (ed.), *To Resist Is to Win!: The Autobiography of Xanana Gusmão with Selected Letters & Speeches,* Melbourne: Aurora/David Lovell Publishing, 2000, p. 39.

49 Hamish McDonald, 'Australia supports Indonesia takeover of East Timor', *National Times,* 15–20 December 1975.

Chapter 4

1. McDonald, 'Indonesia cool on Timor oil search'.
2. Oceanic, 10 December 1975, DFA, *Law of the Sea – Delimitation – Australia – Indonesia (Timor)*, NAA barcode: 1872072, A1838 1733/3/2 Part 7.
3. Woolcott to Canberra, 'Law of the Sea: Indonesia Mochtar', 25 March 1976, DFA, *Law of the sea – Delimitation – Australia – Indonesia (Timor)*, NAA barcode: 1872067, A1838 1733/3/2 Part 2.
4. Record of conversation, Peacock and Mochtar, 15 April 1976, NAA: A1838 1733/3/2 Part 2.
5. Woolcott to Canberra, 16 April 1976, DFA, *Jakarta – [Portuguese] East Timor*, NAA barcode: 4151610, A10463 801/13/11/1 Part 22.
6. Clinton Fernandes, 'Recognition as a political act: Political considerations in recognising Indonesia's annexation of East Timor' in Damien Kingsbury & Costas Laoutides (eds), *Territorial Separatism in Global Politics: Causes, Outcomes and Resolution*, England: Routledge, 2015, p. 98.
7. Record of conversation, Hogue and General Panggabean, Minister of Defence, 23 January 1978, 'Reactions to Australian recognition of East Timor', DFA, Jakarta – East Timor, NAA barcode: 4185704, A10463 801/13/11/1 Part 29.
8. Parsons to Peacock, 'East Timor – Prime Minister's visit', 11 August 1976, *East Timor – Prime Minister's visit to Indonesia – October 1976*, NAA barcode: 4151646, A11443 9.
9. US Ambassador to Indonesia, Newsom to US Embassy Canberra, 13 October 1976, Wikileaks, https://wikileaks.org/plusd/cables/1976JAKART13384_b.html.
10. Alfred Parsons, *South East Asian Days*, Nathan, Qld.: Centre for the Study of Australia–Asia Relations, Faculty of International Business and Politics, Griffith University, 1998, p. 139.
11. Michael Richardson, 'US warns Australia on Timor situation', *Sydney Morning Herald*, 3 August 1976.
12. 'Now for talks on seabed rights' *Australian*, 9 October 1976.
13. Gough Whitlam, 'Motion of want of confidence', *Hansard*, House of Representatives, 12 October 1976.
14. Malcolm Fraser, 'Motion of want of confidence', *Hansard*, House of Representatives, 12 October 1976.

15 'PM accused of "illegal" talks on sea border', *Canberra Times*, 18 October 1976.
16 Mike Steketee, 'Seabed border plan shelved', *Canberra Times*, 19 October 1976.
17 Michael Richardson, 'Indonesia's Timor carrot', *Australian Financial Review*, 19 October 1976.
18 Brazil to Attorney-General, 20 October 1976, Attorney-General's Department, *Portuguese Timor – Continental Shelf*, NAA barcode: 11089376, A5034 SR1974/3009 Part 2.
19 Fernandes, 'Accomplice to mass atrocities', p. 6.
20 DFA, *Jakarta – Timor – Dunn allegations – Australian reactions – Australian parliamentary activity*, NAA barcode: 4209502, A10463 801/13/11/10 Part 1; and DFA, *Jakarta – Third country relations – East Timor – Dunn allegations*, NAA barcode: 4227149, A10463 801/13/11/10 Part 2.
21 James Dunn, interview with author, Canberra, 21 September 2016.
22 Russell Skelton, 'Indons killed 60,000 report', *Age*, 19 November 1976; Bruce Juddery, 'Estimate of 100,000 killed in E. Timor', *Canberra Times*, 20 November 1976; James Dunn, *East Timor: A Rough Passage to Independence*, Double Bay, NSW: Longueville Books, 2003, p. 268.
23 James Dunn, 'Situation in East Timor', November 1976, DFA, *Jakarta – Portuguese Timor – General*, NAA barcode: 4185639, A10463 801/13/11/1 Part 27.
24 Fretilin media statement, 28 October 1976, NAA: A10463 801/13/11/1 Part 27.
25 Tom Allard, '"Sounds like fun": Aussie diplomats mocked reports of Indonesian rape and murder of Timorese', *Sydney Morning Herald*, 21 February 2016.
26 Taylor to DFA Canberra, 10 December 1976, NAA: A10463 801/13/11/1 Part 27.
27 Woolcott to DFA Canberra, 14 December 1976, NAA: A10463 801/13/11/1 Part 27.
28 DFA Canberra to Peacock, 25 February 1977, NAA: A1838 1733/3/3 Part 2.
29 Taylor record of conversation with British High Commission, 17 February 1977, DFA, *Indonesia – Political Relations with Australia – General*, NAA barcode: 1509117, A1838 3034/10/1 Part 48.

30	James Dunn, 'The East Timor situation – Report on talks with Timorese refugees in Portugal', NAA: A10463 801/13/11/10 Part 2.
31	Ben Kiernan, 'Cover-up and denial of genocide: Australia, the USA, East Timor and the Aborigines', *Critical Asian Studies*, vol. 34, no. 2, 2002, pp. 171–2.
32	Michael Hodgman, 'Timor appeasement must end', *Australian*, 21 February 1977.
33	Tony Walker, 'Indon threat on Timor, Canberra told "Keep Dunn quiet"', *Age*, 16 March 1977.
34	Transcript, Parliament Joint Committee on Foreign Affairs and Defence Sub-Committee on Territorial Boundaries, (Timor Boundaries) 30 March 1977, in camera, NAA: A1838 1733/3/2 Part 2.
35	AAP, 'East Timor dead, "50,000"', *Canberra Times*, 1 April 1977.
36	Leach to Gordon, 4 April 1977, NAA: A1838 1733/3/3 Part 2.
37	'East Timor: Visit by Hogue and Rogers', 26–30 April 1977, DFA, *Jakarta – Third country relations – East Timor – Visits to East Timor,* NAA barcode: 4227150, A10463 801/13/11/11 Part 1.
38	Hogue to Secretary DFA, 'East Timor: Visit by Pro-Numcio', 29 June 1977, NAA: A10463 801/13/11/10 Part 2.
39	Yellow envelope, DFA, *Jakarta – Portuguese Timor – Press,* NAA barcode: 4185994, A10463 801/13/11/3 Part 5.
40	Michael Richardson, '"Jakarta's rule a fact"', *Age*, 16 June 1977.
41	Woolcott to Parkinson, 19 May 1977, NAA: A10463 801/13/11/10 Part 2.
42	Parkinson to Woolcott, 24 May 1977, DFA, *Jakarta – East Timor,* NAA barcode: 4185638, A10463 801/13/11/1 Part 28.
43	Michael Richardson, *Australian Financial Review*, 25 May 1977.
44	J.F. Dubourdieu, Managing Director, Australian Aquitaine Petroleum, to the Delegate of the Designated Authority Department of the Northern Territory, 26 May 1977, NAA: A1838 1733/3/3 Part 2.
45	Minister for Natural Resources Doug Anthony to Acting Minister for Foreign Affairs Ian Sinclair, 7 June 1977, NAA: A1838 1733/3/3 Part 2.
46	Sinclair to Anthony, 23 June 1977, NAA: A1838 1733/3/3 Part 2.

47 Woodside to the Delegate of the Designated Authority, Mines Branch, Department of the Northern Territory, 29 June 1977, NAA: A1838 1733/3/2 Part 2.
48 DFA Canberra to Jakarta, East Timor: Dowd Report, 21 July 1977, NAA: A10463 801/13/11/10 Part 2.
49 Carleton to Woolcott, 19 July 1977, NAA: A10463 801/13/11/1 Part 28.
50 AAP, 'Indonesians shot 150 on Dili pier', *Age*, 2 August 1977; Richard Carleton, 'Inside the island of agony: Timor and a story of a massacre', *Age*, 10 August 1977; Richard Carleton, 'The place of death in Timor', *Age*, 11 August 1977.
51 Duggan, 'Note for file Richard Carleton's articles', 17 October 1977, National Archives UK: 8777886 FCO 15/2252.
52 Woolcott to Allen and Hogue, 5 September 1977, NAA: 28 A10463 801/13/11/1 Part 28.
53 Woolcott to Allen and Hogue, 31 October 1977, NAA: 28 A10463 801/13/11/1 Part 28.
54 Woolcott to Parkinson, 7 December 1977, 'Timor', 7 December 1977, DFA, 'Jakarta – East Timor', NAA: A10463 801/13/11/1 Part 29..

CHAPTER 5

1 Department of State to Commander in Chief US Pacific Command, 3 January 1978, Wikileaks, https://www.wikileaks.org/plusd/cables/1978STATE000345_d.html.
2 David Goldsworthy et al., *Facing North: A Century of Australian Engagement with Asia, Volume 2 – 1970s to 2000*, Carlton: Melbourne University Press and DFAT, 2003. p. 217.
3 Fernandes, 'Recognition as a political act: Political considerations in recognising Indonesia's annexation of East Timor', p. 100.
4 Andrew Peacock, Minister for Foreign Affairs, 'Submission No. 1865 – East Timor – Australia Policy, Foreign Affairs and Defence Committee', 11 January 1978, Cabinet Office, *Submission No 1865: East Timor – Australian Policy – Decision 4485(FAD)*, NAA barcode: 8911897, A12909 1865.

5	In addition to the paragraph quoted, the redacted text is in sections dealing with 'other important issues', the recognition issue and 'timing'.
6	Canberra to NY, 4 October 1978, DFA, *Law of the Sea – Delimitation – Australia – Indonesia (Timor)*, NAA barcode: 1872068, A1838 1733/3/2 Part 3.
7	Woolcott to Mochtar, 20 January 1978, NAA: A10463 801/13/11/1 Part 29.
8	Woolcott to Secretary DFA, 'Minister's statement on Timor', 20 January 1978, NAA: A10463 801/13/11/1 Part 29.
9	Draft Delimitation of Sea and Seabed Boundaries, April 1978, DFA, *Law of the Sea Delimitation – Australia – Indonesia (Timor)*, NAA barcode: 1872068, A1838 1733/3/2 Part 3.
10	Peter Terry, 'Way opens for Timor Oil hunt', *Australian Financial Review*, 21 February 1978, quoted in King, 'A gap in the relationship', p. 20.
11	Record of conversation, Ambassador Critchley, Embassy First Secretary, David Irvine and Indonesian Brigadier-General Adenan, Director-General for Foreign Relations and Security in Indonesia's Department of Foreign Affairs, 24 May 1978, DFA, *Indonesia – Relations with Portuguese Timor*, NAA barcode: 1500806, A1838 3006/4/3 Part 24. Irvine was director deneral of ASIS when the alleged bugging operation took place in 2004.
12	Quoted in Fernandes, 'Accomplice to mass atrocities', p. 7.
13	Smith to Peacock, 26 April 1978, NAA: A1838 1733/3/2 Part 3.
14	Smith to Peacock, 26 April 1978.
15	DFA speaking notes, 27 July 1978, NAA: A1838 1733/3/2 Part 3.
16	Cabinet Office, *Submission No 2472: Delimitation of the Australian continental shelf and 200 Nautical Mile Fishing Zone – Decision 6156*, NAA barcode: 8147460, A12909, 2472.
17	Jakarta to Canberra, 17 July 1978, NAA: A1838 3006/4/3 Part 24.
18	Alexander to Canberra, 'East Timor', 15 August 1978, DFA, *Jakarta – East Timor*, NAA barcode: 4185703, A10463 801/13/11/1 Part 30; and see commentary in Fernandes, 'Accomplice to mass atrocities', p. 7.

19	Douglas Campbell, interview with author, Canberra, 19 December 2014.
20	Parker & Parker to Peacock, 4 September 1978, NAA: A1838 1733/3/2 Part 3.
21	Quoted in Arnold Kohen & Roberta Quance, 'The politics of starvation', *Inquiry*, 18 February 1980, p. 20.
22	Parsons to Peacock, 'East Timor – Prime Minister's visit', 11 August 1976, NAA: A11443 9.
23	Richard Gill, 'Out of the hills comes Timor's tragedy', *Herald*, 11 September 1978; Richard Gill, 'Disaster around idyllic Dili', *Herald*, 12 September 1978.
24	David Jenkins, 'Timor's arithmetic of despair', *Far Eastern Economic Review*, 29 September 1978.
25	Peter Hastings, *Sydney Morning Herald*, 11 September 1978.
26	Woolcott to Parkinson, 'Timor', 7 December 1977, NAA barcode: A10463 801/13/11/1 Part 29.
27	Desmond Ball & Hamish McDonald, *Death in Balibo, Lies in Canberra*, St Leonards, NSW: Allen & Unwin, 2000; Jim Aubrey, 'Canberra: Jakarta's Trojan Horse in East Timor' in Stephen McCloskey & Paul Hainsworth (eds), *The East Timor Question: The Struggle for Independence from Indonesia*, I.B. Tauris, 2000, p. 140; Fernandes, 'Accomplice to mass atrocities'.
28	Kohen & Quance, 'The politics of starvation'.
29	Kohen & Quance, 'The politics of starvation', p. 20.
30	Woodside to Delegate of the Designated Authority, 18 July 1978, NAA: A1838 1733/3/2 Part 3.
31	Department of Defence, Captain E.E. Johnston (Australian Delegation) and Lieutenant-Colonel Yahya (Indonesian Delegation), 12 September 1978, NAA: A452 1972/2593.
32	Draft record of conversation, Mochtar and Peacock, New York, 5 October 1978, NAA: A10463 801/13/11/1 Part 30.
33	Draft record of conversation, Mochtar and Peacock.
34	Niner, *Xanana*, p. 34.
35	Douglas Campbell, 'Report of visit to East Timor', 13 November 1978, DFA, Jakarta, NAA Correspondence Files, 1978.
36	Douglas Campbell, interview with author, Canberra, 19 December 2014.
37	Cabinet Decision 7157, referred to in draft brief for Minister for discussions with Indonesian foreign minister

delimitation of maritime boundaries, 6 December 1978, NAA: A1838 1733/3/2 Part 5.

38 Niner et al., *To Resist Is to Win!*, p. 57.

39 'Extracts relating to seabed boundary negotiations from the press conference given by the Minister for Foreign Affairs, Mr Andrew Peacock, on 15 December 1975, during the visit to Australia of the Indonesian Foreign Minister, Professor Mochtar Kusumaatmadja', NAA: A1838 1733/3/2 Part 5.

40 Clinton Fernandes, *The Independence of East Timor: Multi-Dimensional Perspectives – Occupation, Resistance, and International Political Activism*, The Sussex Library of Asian Studies, Brighton, England : Sussex Academic Press, 2011, p. 55.

41 ABC News, 2 January 1978, handwritten transcript, NAA: A1838 3006/4/3 Part 24.

42 Cabinet Office, *Cabinet Submission 618 – Australia-Indonesia maritime delimitation negotiations, Decision 2734*, 9 January 1984, NAA barcode: 31424609, A13977, 618.

43 'Petroleum Prospectivity of the "Timor Gap Area"', Department of National Development, *Law of the Sea – Delimitation – Australia – Indonesia (Timor)*, 1 May 1979, NAA barcode: 1763272 A1838 1733/3/2 Part 9.

44 *Chega!: The Report of the Commission for Reception, Truth and Reconciliation Timor-Leste*, Dili: CARV, 2005 p. 64.

45 Peter Rodgers, 'Horror on our doorstep', *Age*, 1 November 1979, and 'Agony on our doorstep', *Sydney Morning Herald*, 1 November 1979.

46 A week later Mochtar admitted that only half of the pre-1975 population of East Timor was under Indonesian control, and that 120,000 people had died since the 'civil war' began in 1975: John Taylor, *Indonesia's Forgotten War: The Hidden History of East Timor*, Leichhardt, NSW: Pluto Press, 1991, p. 203.

47 Niner, *Xanana*, pp. 79–80.

48 *Cabinet Submission 12 – Australian policy on Indonesia – East Timor – Decisions 93/DER and 116*, Prime Minister's Department, NAA barcode: 31405612, A13977 12.

49 José Ramos-Horta, *Funu: the Unfinished Saga of East Timor*, Trenton, N.J.: Red Sea Press, 1987, p. 83.

50 *Chega!*, pp. 66 & 120.

51 Chinkin, quoted in Aubrey, 'Canberra', p. 142.
52 *Cabinet Submission 618,* NAA: A13977 618.
53 Senate Foreign Affairs, Defence and Trade References Committee, *Final Report on the Inquiry into East Timor,* Australian Parliament, 2000, p. 160.
54 Niner, *Xanana,* p. 105.
55 Niner et al., *To Resist Is to Win!,* p. 119.
56 Gusmão to McIntosh, 19 April 1988, https://timorarchives.wordpress.com/2016/05/01/mcintosh-ulun-toos/.
57 Mark Aarons & Robert Domm, *East Timor: A Western Made Tragedy,* Sydney: Left Book Club, 1982, p. 27. The interview aired on the ABC's *Background Briefing* in October 1991.
58 This treaty has not been ratified by the Indonesian parliament, which does not want to be reminded just how much territory Indonesia conceded to Australia in 1972.

CHAPTER 6

1 John Howard, 'Foreword', in John Blaxland, *East Timor, A retrospective on INTERFET,* Carlton: Melbourne University Publishing, 2015, p. 7.
2 Department of Foreign Affairs and Trade, *East Timor in Transition, 1998–2000: An Australian Policy Challenge,* Canberra: DFAT, 2001.
3 Clinton Fernandes, *Reluctant Saviour: Australia, Indonesia and the Independence of East Timor,* Carlton North, Vic.: Scribe Publications, 2004, pp. 1–2.
4 Mark Baker, 'Downer "verges on racism", says Horta', *Sunday Age,* 16 August 1998.
5 Geoffrey McKee, 'The new Timor Gap: Will Australia now break with the past?' in Lansell Taudevin & Jefferson Lee (eds), *East Timor: Making Amends?: Analysing Australia's Role in Reconstructing East Timor,* Otford, NSW: Australia–East Timor Association, Otford Press, 2000, p. 100.
6 Beanland, 'East Timor & the politics of oil'.
7 Joshua Frydenberg & Greg Hunt, 'Timor plan is more palatable after dinner', *Australian,* 14 January 1999. Hunt was an adviser to Downer from 1994 to 1998 and Frydenberg from 1999 to 2003.

8	Richard Woolcott, 'It's time to recall a treaty', *Australian Financial Review*, 19 January 1999.
9	Fernandes, *Reluctant Saviour*, p. 3.
10	Laurie Brereton, 'East Timor: Selective and partisan publication of DFAT records', News release, 17 July 2001.
11	Downer, Hansard, House of Representatives, 12 October 1999.
12	Tiffen, *Diplomatic Deceits*, p. 86.
13	Geoffrey Robinson, *East Timor 1999 Crimes against Humanity*, Report commissioned by the United Nations Office of the High Commissioner for Human Rights, July 2003.
14	Niner, *Xanana*, p. 191.
15	Fernandes, *Reluctant Saviour*, pp. 56–57.
16	A.S. Blunn, *Report of the Inquiry of Mr A.S. Blunn AO on Behalf of the Inspector-General of Intelligence and Security into the Investigation into Alleged Security Breaches by the Late Mervyn Jenkins*, para. 121, https://www.igis.gov.au/sites/default/files/files/Inquiries/docs/Jenkins_Final_Report_2000.pdf. See also http://www.abc.net.au/4corners/mervjenkins/blunnreport/blunn2.htm.
17	Senate Foreign Affairs, Defence and Trade References Committee, *Final Report on the Inquiry into East Timor*, p. 62.
18	Niner, *Xanana*, p. 206.
19	Alexander Munton, 'A study of the offshore petroleum negotiations between Australia, the UN and East Timor', Thesis, Doctor of Philosophy, Australian National University, 2006, p. 99.
20	Munton, 'A study', p. 99.
21	Niner, *Xanana*, p. 209.
22	Senate Foreign Affairs, Defence and Trade References Committee, *Final Report on the Inquiry into East Timor*, p. 63.
23	Jonathan Morrow & Rachel White, 'The United Nations in transitional East Timor: International standards and the reality of governance', *Australian Year Book of International Law*, vol. 22, 2002, p. 25.
24	Munton, 'A study', p. 15.
25	Cleary, *Shakedown*, p. 57.

26 Mark Baker, 'Australia "pocketing" Timor royalties', *Sydney Morning Herald*, 19 November 2003.
27 Cleary, *Shakedown*, p. 19.
28 Alexander Downer and Mari Alkatiri, 'Timor Sea Treaty ministerial meeting', *Crikey*, 27 November 2002.
29 Cleary, *Shakedown*, p. 82.
30 Cynthia Banham, 'Fair play demanded in oil talks', *Sydney Morning Herald*, 11 December 2003.
31 Bernard Collaery, 'National security, legal professional privilege, and the Bar Rules', Address at the Australian National University, 11 June 2015, https://law.anu.edu.au/sites/all/files/events/national_security_legal_professional_privilege_and_the_bar_rules_print.pdf.
32 Paul Cleary, 'The 40 year battle over Timor's oil', *Australian*, 5 December 2013.
33 Mark Aarons, 'Oil and water: Australia blurs the lines with Timor-Leste', *Monthly*, December 2015.
34 'Moraitis firmly made clear, yet again, that Australia was completely unwilling to look at any boundary proposal outside the JPDA': Cleary, *Sydney Morning Herald*, p. 179.
35 Gleeson, Opening Session, Permanent Court of Arbitration Case No 2016-10, p. 92.

Conclusion

1 'East Timor spying scandal: Tony Abbott defends ASIO raids on lawyer Bernard Collaery's offices', ABC News, 4 December 2013, http://www.abc.net.au/news/2013-12-04/asio-arrests-key-witness-in-east-timor-spying-scandal/5132954.
2 Xanana Gusmão, *Timor-Leste's Story: Securing Its Sovereignty over Land and Sea*, New York: International Peace Institute, 1 October 2015.

A Note on the National Archives of Australia

1 National Archives of Australia to author, 14 April 2015.
2 Email from Australian Government Solicitor to author, 21 October 2016.

www.ingramcontent.com/pod-product-compliance
Lightning Source LLC
Chambersburg PA
CBHW050524170426
43201CB00013B/2077